TREADING ON THIN AIR

TREADING ON THIN AIR

Atmospheric Physics, Forensic Meteorology, and Climate Change: How Weather Shapes Our Everyday Lives

ELIZABETH AUSTIN, PH.D.

PEGASUS BOOKS
NEW YORK LONDON

TREADING ON THIN AIR

Pegasus Books, Ltd.
148 West 37th Street, 13th Floor
New York, NY 10018

Interior design by Maria Fernandez

Library of Congress Cataloging-in-Publication Data is available.

ISBN: 978-1-68177-403-9

Printed in the United States of America

For my family . . .

My loving and supportive husband, Alan Austin
Our son, nuestro Angelito, Evan
My sister, Greer, and brother, Patrick
And my parents, Catherine and Patrick Williams

CONTENTS

TREADING ON THIN AIR

Author's Note

I wrote this book from my memories, notes, articles, and research. As some of the events go back many decades, my memories have faded. I tried to recall all people, names, and facts to the best of my knowledge. I did not change any names, and all events are true to my recollections. This book is not intended to be a purely scientific book nor a standard memoir but a unique combination to give the reader a journey through the world of weather and climate through my eyes.

ONE
Surfing in the Stratosphere

I think the age of exploration is just beginning, not ending, on our planet.

—Robert Ballard

"What's the most dangerous thing you have ever done, Einar?" the reporter asked. "Was it flying a jet fighter in the U.S. Air Force, as a test pilot for NASA, for Grob, for the Royal Air Force, or was it trying to set the world altitude record in the Perlan glider with Steve Fossett?" Without hesitating, pilot Einar Enevoldson replied, delivering his response in an incredibly understated voice, as if he were making an observation about a move in a chess match, "Oh, flying the Perlan to altitude," he said, as if he were merely summarizing what he had eaten for lunch that day. I've found that most pilots of this caliber usually don't have a shred of Hollywood melodrama about themselves when it comes to describing their incredible, often extremely dangerous, feats.

It was a dark and very cold winter morning in August 2002 in the small town of Omarama, New Zealand. Omarama is on the South Island of New Zealand and is in the lee of the Southern Alps in central Otago province. Omarama, which in Māori means "place of light," has some of the clearest skies I have ever seen. Omarama is so small, it's almost *not* a town. The population is just over two hundred souls and, on this occasion, our Perlan team had essentially taken over the place. The team members were buzzing around the one grass runway at the airport, busy making all the preparations in an attempt to break the world altitude record for a manned glider. The press was everywhere with cameras rolling as they interviewed various people from our team. The locals were chatting away and some volunteers were doing their best to keep the people and press in their "designated areas." I was one of the team members giving interviews to the media as well, as I was the chief meteorologist on the project. Television production company Natural History New Zealand (NHNZ) got wind of the project and found it interesting enough to come and do an on-camera interview with me, too. Little did most of the interviewers know that our team had already spent years getting to this point in the project, a fact that would soon be revealed, however, as they got their stories out of various members of the team.

The glider built for the Perlan mission is special, as it does not have an engine and, unlike most gliders, is designed for high-altitude flight. Some gliders carry one pilot and others two pilots. Our Perlan glider carries two pilots. It has a tremendously long wingspan in order to facilitate flight to very high altitudes. Gliders are also inherently strong, as they must be able to withstand tremendous turbulence when flown in mountainous regions where turbulence can hide in plain sight, so to speak. The Perlan Glider is built to withstand 8 g's or eight times the force of gravity at sea level.

The locals turned out to see what the crazy Americans were doing in the wintertime soaring a glider in the mountains, and most of the press were there because Steve Fossett, the billionaire adventurer, was there to attempt another world record on the heels of his becoming the first person to fly around the world alone, nonstop in a balloon, covering 19,428 miles.

All of this commotion outside the glider was in stark contrast to the scene inside the cockpit of the experimental craft as Steve Fossett and Einar Enevoldson, who would be piloting the glider together, calmly completed their checklists. During the preflight time it was critical that they remained calm, as they didn't want to break a sweat in their space suits. This could have spelled disaster during the flight, since something as simple as perspiration could fog up their facemasks, making it impossible for them to see.

"Checklist complete," Einar said to Steve.

"Roger that," Steve replied through the microphone in the helmet of his pressure suit.

Mike Todd, our team's life support specialist, who was formerly with NASA, reached over to plug the glider's oxygen system into the pressure suits and then disconnect the external oxygen tanks the pilots had been using to breathe for the previous hour. Mike is an expert skydiver and parachute jumper who worked in the High Altitude Life Support and Pressure Suit Division of Lockheed's Skunk Works for almost thirty years.

Einar and Steve had been pre-breathing pure oxygen for the previous hour in their NASA/Air Force pressure suits. They are just like the space suits astronauts wear. Pre-breathing pure oxygen is a necessary protocol, whether you are about to perform a space walk outside the International Space Station or take a flight to high altitudes in an unpressurized aircraft while wearing these suits—just as Einar and Steve were about to do. Pre-breathing pure oxygen washes out any excess nitrogen in the body prior to the flight. At altitudes above 50,000 feet a person, even one wearing a space suit like the ones Einar and Steve were wearing, requires pressurization because the atmospheric pressure is so low. In fact, above 55,000 feet a person's blood will literally "boil" if he or she does not have a pressure suit: low pressure changes the equilibrium point—the point at which gas bubbles form in liquids, including blood. A rapid drop in pressure can result in caisson disease, known as the bends among the diving community. Caisson disease is the formation and expansion of nitrogen gas bubbles in the bloodstream when a person is exposed to a rapid drop in external pressure—which was precisely what Steve and Einar were going to be exposed to while making this record attempt. Wearing the pressure suit enables the pilots to breathe by providing the necessary pressure (lungs don't function without positive pressure—which is also why aircraft are pressurized), and breathing pure oxygen helps flush out the nitrogen that builds up in the blood.

"All the gases in the blood start to come out of solution and turn back into gas rather than a liquid. Essentially, your blood just turns to foam. You don't work too well like that," Mike Todd explained to the press.

The Perlan phase 1 glider cabin was not pressurized, hence the need for pressurized suits. This phase 1 glider was a highly modified German Glaser-Dirks DG505M (without the motor) sailplane and categorized in the experimental class of aircraft. The pressure suits keep the pilots' bodies pressurized as if they were at 35,000 feet. But during this flight they will be attempting to fly above 50,000 feet. For comparison, Mount Everest's peak is at 29,029 feet, and to put this into perspective, most commercial airliners cruise in the mid-30,000-foot altitude range. The engine of the glider had been removed and life support systems, liquid oxygen containers, and instruments had been added in its place.

The Perlan I glider was a highly modified, off-the-shelf glider, whereas the Perlan II glider has been built from scratch. This experimental Perlan II glider has an 84-foot wingspan and a gross weight of 1,800 pounds. The cabin is partially pressurized at 8.5 pounds per square inch differential. This means that at about 39,000 feet, the glider cabin can maintain pressure equivalent to 8,000 feet, and as they fly higher the cabin pressure also goes higher. A conventional aircraft with engines cannot soar in these mountain waves as a glider does but just blasts through them. A glider actually rides these waves upward like an elevator. A glider can also be maneuvered up and downwind to locate regions of lift in the lee (downwind) side of mountains, making it a perfect platform for scientific instruments and measurements.

But what is life like that high up in the sky? Quoting a note from Einar, as a bit of tragicomic relief, he paints a vivid picture of the incredibly hostile environment at high altitude, while also reinforces the pluck of these pilots who venture up into these elements.

> *Thirty-nine thousand feet is the limit for 100% oxygen, no leaks, perfect mask. One can then "pressure breathe" where the lungs are inflated above ambient pressure, to keep the pressure in the lungs at 39,000 feet. You must force the air out to exhale—it gets to be a lot of work eventually. I have spent a couple of hours at 43,000 feet in the old days in USAF fighters with poorly operating pressurization systems. It also invites "the bends." In extreme necessity one can go to about 46,000 feet for a few minutes, but hypoxia is setting in, and ten minutes or so is about the limit. Exhaling is very hard and tiring. The oxygen mask is painfully tight, and must be very precisely fitted. It is common to blow a lot of air back out through the tear ducts. It makes your eyes water. Then if your masochistic nature takes over, you can put on a "pressure jerkin"—a heavy shirt lined with a bladder that inflates. This will counter the internal pressure in the lungs and make breathing easier. There is a catch—your blood is forced away from the pressure, into your lower abdomen, and legs, feet, and hands. So you put on a G-suit—pants with bladders over your abdomen, thighs, and calves. You will still faint in a few minutes because blood still finds the unpressurized nooks and crannies, and also just squeezes out of your veins into surrounding tissue. Low blood volume causes the fainting. By balancing the oxygen pressure just right, you can pass out from hypoxia, and faint at the same time. This is logic of the "pressure jerkin" system. You can depressure at 85,000 feet, then descend to 40,000 feet in less than four minutes, and be conscious and functional the whole time. Bill Dana (another NASA test pilot) and I tested it in an F-104 from 60,000 feet. We took turns depressurizing, with one guy in a full pressure suit and one in the jerkin flying the airplane. Worked okay!*

Sounds like fun, right?

Malcolm Walls, our tow plane pilot, was sitting in the single-engine Piper Super Cub that would tow us off the ground, waiting for his signal to go. He got the thumbs-up and they were off. Malcolm towed the Perlan I glider behind the Super Cub up to around 9,000 feet, where the glider was released from the tow plane. Einar and Steve were now soaring on their own as they began searching for the best lift that would take them even higher and, they hoped, up into the stratosphere.

Einar Enevoldson has been flying gliders since he was very young, and he feels his way as much as he looks for the lenticular (lens-shaped) clouds that serve as clues where to find lift. Sometimes there are no clouds to provide clues to the pilots and they must rely solely on the glider's instruments and their knowledge of where lift occurs. Gliders fly on lift and the air currents moving up and down. The altimeter, an instrument that shows the altitude, does not respond quickly enough for a pilot to use it to find and remain in regions of lift. So they use an additional instrument called a variometer. A variometer responds instantly and gives a beeping sound that gets higher and faster as the lift increases: the pilot just listens in order to find lift. The sound falls in tone if the variometer detects descending air to let the pilot know that the glider is sinking.

This Omarama flight represents only phase one of a three-phase project with gliders, and we have a long way to go. Each subsequent phase is exponentially more difficult than the last. The Perlan Project was and still is truly an amazing adventure with many benefits, including providing a scientific platform for researching areas of our atmosphere such as the tremendous turbulence of breaking stratospheric mountain waves, how the energy of these waves is dispersed, and if these cause injections of energy and chemical constituents between the stratosphere and troposphere layers of the atmosphere, which will have a tremendous impact by improving our weather forecasting and climate models. But there are also many risks: the glider flying into a region of these breaking waves could break apart or have a rapid decompression of the cabin at high altitudes, either of which would spell disaster and perhaps result in the deaths of the pilots.

Once Einar and Steve achieved Phase I, we literally set our sights higher. Phase II, currently taking place, is to reach 90,000 feet, and then Phase III—the ultimate goal—is to soar to 100,000 feet (30 kilometers), which would bring the glider near the edge of space. At 100,000 feet, the pressure of the atmosphere is so thin, it is similar to the pressure on the surface of Mars. The behavior of the atmosphere at this level is of particular interest to the aviation community—engineers, aerodynamicists, and manufacturers of aircraft—as it will give insight into flight on Mars.

Perlan is an Icelandic word meaning "pearl." Perlan is the name given to this project, inspired by mother-of-pearl or nacreous clouds occasionally seen at high altitudes and high latitudes. The research that we will be conducting will help to fill in many gaps in our scientific knowledge and help to improve our atmospheric models for forecasting weather and for climate predictions. We collect data from instruments mounted on the glider, from ground-based instruments, and from balloons launched from the ground that travel up through the atmosphere, but the balloons cannot be guided through particular regions as the glider can. The fact that we can fly slowly through these stratospheric mountain wave structures, flying transects or choosing to fly vertically straight upward during our research will help answer some major questions in the research of stratospheric mountain waves. Standard aviation measuring techniques cannot measure things like mountain wave dynamics, breaking waves, and multi-scale frontogenesis diagnostics (a fancy term for how the horizontal temperature gradients tighten to produce weather fronts, like cold and warm fronts), but a glider can. The project will help us fully understand the role of mountain wave dynamics in vertical exchanges between the troposphere and the stratosphere.

Back down in New Zealand, the world altitude record we were trying to beat was 49,009 feet set back in 1986 by Robert "Bob" Harris on a flight in the Sierra Nevada mountains of California. Alan Austin—my husband—and I have become friends with Bob over the years, and Bob and I still discuss the unique meteorological conditions it took for him to reach almost fifty thousand feet in a glider. Bob revealed in great detail how he spent years watching the weather, going to his local National Weather Service office (as this was before the Internet days), and talking to the forecasters about the weather until one day, he said, "it was perfect." A perfect day for a glider flight of this type is one with a stable lower atmosphere near mountain or ridgetop level with less stability aloft; prefrontal conditions—when there is an approaching frontal system (postfrontal are sometimes good, too) such that there was a cold front approaching the West Coast of the United States; winds of at least 20 knots or greater at ridgetop; winds hitting the Sierra Nevada mountain range at an angle nearly perpendicular to the ridgeline (which runs northwest to southeast, meaning winds hitting the mountains in the northeast-southwest direction); and the location of the jet stream over the region such that the winds were increasing smoothly with altitude with very little directional change (i.e., no wind shear). The Sierra Nevada region of California is not impacted by the polar vortex, so getting to extremely high altitudes (65,000 feet or higher) is not possible in a glider.

In New Zealand, however, that day, our Perlan glider was going to need considerably more specific and critical weather conditions—namely, in addition to all of the above criteria, we needed to be on the edge of the polar vortex, providing increasing winds with altitude all the way up through the stratosphere, occurring all at once in order to enable setting a world record.

On that day in Omarama, however, a new world record was not to be. Why did we trek all the way down to Omarama? It is known for its world-class gliding in the summertime. But we had to be there in the southern winter, as that is the time when the polar vortex actively allows mountain air waves to travel all the way from where we live in the troposphere up into the stratosphere. Though the press mostly butchers the explanation of its characteristics and what its effects are, the polar vortex is truly amazing.

The polar vortex is a system of cold, low-pressure air circulating around each pole. It exists in the northern hemisphere usually between the months of November through April and then dies out. In the southern hemisphere, however, the polar vortex forms during June and dies out sometime between October and November. The vortex covers a large-scale area from the polar regions up to around 45 to 50 degrees latitude north and south, respectively, during the winter seasons. Vertically, it extends from near the ground, where we live and breathe, up to the troposphere, and then into the stratosphere, and from there on up through the stratosphere into the layer above, called the mesosphere. The polar vortex can extend upward 30 miles (50 kilometers).

The edge of the vortex is surrounded by a jet stream, called the *polar night jet*, which is similar to the jet streams we might be familiar with from commercial airline flights. Those familiar jet streams exist around 30,000 feet above us, but the polar night jet exists at much higher altitudes—extending from the tropopause region which (where we currently stage our world record flights out of in El Calafate, Argentina) is near the poles at about 10 km (33,000 feet). Though the stratosphere extends up to 50 km (165,000 feet), the polar night jet continues on through the entire stratosphere and into the mesosphere, up well beyond 165,000 feet (over 30 miles). The center, or peak winds, of the polar night jet generally occur around 120,000 to 125,000 feet up in the atmosphere over the southern hemisphere. These winds develop along a zone of sharp temperature contrasts that occur during the long, cold polar nights. In this zone, the stratosphere cools down by radiating its heat to space, similar to how the surface of the Earth cools during a clear, calm night. In this case, however, the cooling takes several months, unlike the overnight cooling of the Earth. The polar night jet associated with the polar vortex is located in the upper regions of the troposphere, up in the stratosphere, and into lower mesosphere regions. This jet can attain speeds up to 300 miles per hour (260 knots). For comparison, a typical jet stream that commercial aircraft use to get across the United States at around 30,000 feet flows at about 80 to 140 miles per hour.

There are actually subtropical and polar jet streams that impact the midlatitudes (including the continental United States) at different times of year. During the winter months, the polar jet stream tends to impact these regions; during the summer months, the subtropical jet moves north, impacting these regions. This situation is mirrored in the southern hemisphere. The polar jet stream is stronger than the subtropical, and can reach 250 to 275 miles per hour, though this is not the norm. This jet stream usually occurs between six and eight miles above sea level (about 30,000 to 42,000 feet), whereas the stronger polar night jet that the Perlan glider requires in order to attain record flight altitudes occurs in the upper troposphere up through the stratosphere and sometimes into the mesosphere, up to altitudes of 165,000 feet.

The polar night jet isolates the air within the polar vortex, which, specifically in the southern hemisphere, exists in the ozone hole over

Antarctica. (In the northern hemisphere, the vortex exists at latitudes of about 50 to 60 degrees north.) The ozone hole is not really a hole but a region of very low ozone concentrations. An ozone hole is defined as a region with 220 Dobson Units or lower of ozone. (Dobson Units describe the compressed thickness of the ozone layer.) The ozone layer in the stratosphere is what protects us humans from ultraviolet radiation, which is responsible for sunburn and skin cancer. In the northern hemisphere we are nearing the definition of a hole in the ozone, too, specifically in Greenland and the northern regions of Finland, Sweden, Norway, Russia, Canada, and Alaska. However, as with the southern hemisphere polar vortex, the northern one wiggles and wobbles and changes shape, sometimes even dipping much farther south to include the northern portion of the contiguous United States. Thus both the ozone hole in the southern hemisphere and the region of ozone depletion in the northern hemisphere move and change shape during their respective seasonal lifetimes. These "holes" are caused by a combination of atmospheric conditions in which the long winter months of darkness and cold over the poles, along with the polar vortex or whirlpool of stratospheric winds, ends up isolating the air in the center.

The cold, high polar stratospheric clouds that form in the arctic during these winter months are composed of water, ice, or nitric acid. On the surface of these particles that make up these clouds, unusual chemical reactions with chlorine or bromine occur, breaking down the ozone. Chlorine and bromine are halogen gases (noble gases). Bromine exists in the earth's crust and also comes from agricultural pesticides and the combustion of leaded gas. Chlorine occurs naturally as it is evaporated from ocean spray, however the majority of chlorine in the atmosphere is from humans through manufactured compounds such as household bleach and also in aerosol sprays. After years of world altitude record attempts in New Zealand, we decided to move the Perlan record attempts to El Calafate, Argentina, another good spot for a glider flight. Though the Andes mountain range is not at its highest down at 50 degrees south latitude, we were 5.5 degrees farther south than we were in New Zealand and that much closer to the edge of the polar vortex. Also, the frontal systems that move through New Zealand during the winter months are sometimes influenced by the

Australian continent, Tasmania, and the Tasman Sea. This changed the orientation of the weather front and impacted the winds so that we didn't get our consistent wind direction of nearly perpendicular to the southern Alps, with altitude, especially around the region between the troposphere and the stratosphere. Argentina does not have this problem, as there is a long fetch of water, the Pacific Ocean, which the frontal systems travel across prior to reaching South America.

The Perlan II flights will be staged out of El Calafate in Argentina because of the more favorable weather conditions that we expect to find there. During those flights, the glider will be soaring on the edge of the ozone hole, allowing us to make important measurements of the atmosphere and its properties, including its chemistry. The Perlan II glider will be equipped with a good deal of scientific measuring equipment and will collect atmospheric and aeronautical data including air data such as winds, air temperatures, ozone, electric and magnetic fields, and concentrations of water vapor and methane. Even though the press and media have recently tumbled onto and adopted the term "polar vortex," they regularly misuse it and incorrectly attribute weather conditions to the polar vortex. I cringe when I hear news or weather readers announce that the weather we are experiencing on any given day at the surface of the earth is *caused* by the polar vortex. The polar vortex wiggles and wobbles as it circulates counterclockwise around the polar regions. Sometimes these wiggles in the northern hemisphere polar vortex extend even as far south as Minnesota, facilitating our lower level jet stream and causing the cold Canadian air to remain over the eastern United States for a longer period of time than other years. The polar vortex does have an effect on certain conditions, but does not directly *cause* the weather, like snow or rain, that the weather readers on television regularly attribute to it.

Our interest in the polar vortex in Argentina had nothing to do with snow and chilly conditions. To soar a glider to 90,000 feet without an engine is only possible with the polar vortex as our ally. To accomplish this feat, we need a weather phenomenon that generates upward moving airflow that the glider can use to climb up to those altitudes, something we call "stratospheric mountain waves."

Mountain waves form whenever an airflow of sufficient speed—at least 20 knots (23 miles per hour) at ridgetops—crosses over a mountain range. Though the taller the mountains the better, the critical component is the polar vortex: we can give up being nearer the taller mountains if we are closer to the edge of the polar vortex. The mechanism behind mountain waves is this: air gets deflected across the top of the mountain range, and as it comes back down, it starts an oscillation that continues downwind of the mountain range. A glider uses the upward-moving part of this oscillation to climb. By remaining in the upward portion of the wave—as long as the wave continues upward with enough energy—the glider will also continue to climb.

These kinds of winds are usually generated by prefrontal conditions (postfrontal conditions are also a possibility in some locations). "Prefrontal" means the time period prior to a frontal wind system moving into the region, e.g., the time before a storm. Generally this time period can range between eight to fifteen hours ahead of the front moving into a given location, depending on how fast the frontal system is moving. So, if we are trying to find ideal wind conditions for a glider, we need to find soaring locations near mountain ranges where the dominant storm systems produce winds roughly perpendicular to the ranges themselves. Sites where this phenomenon occurs include the Sierra Nevada range in California, the Andes in South America, the Southern Alps in New Zealand, the Kebnekaise in Sweden, the European Alps, the mountains in Greenland, the Rockies in the western United States, and the Urals in Russia. Some of these ranges are not close enough to the polar vortex; others are too far north and hence the limited number of daylight flying hours is an issue. Others have too much low-level cloud cover as frontal systems move through: the glider must remain out of clouds for visibility and to avoid icing up. (Flying through visible moisture at below freezing temperatures can cause ice to collect on the aircraft, which degrades the wing surface and increases its weight.) Other mountain ranges, such as the Brook Range in Alaska and the Himalayas, are not oriented as favorably for the necessary perpendicular winds to the mountain range.

The maximum altitude of mountain wave peaks is usually capped at the "tropopause," the boundary between the troposphere and the stratosphere.

This is because the cold air of the mountain wave encounters warmer air at the boundary and cannot rise any farther. The warmer air acts as a fence for these cold mountain waves. On certain occasions, mountain waves can penetrate into the stratosphere when the winds are very strong and there is mixing of the troposphere and the stratosphere. These are called "stratospheric mountain waves." There is a weather phenomenon that generates these "breakthrough" conditions that are ideal for gliders like ours trying to break into high altitudes: the aforementioned—and often maligned—*polar vortex*.

This, along with the fact that there is open water for miles to the west of South America, as opposed to the Tasman Sea and then the continent of Australia to the west of New Zealand, turned out to make all the difference as far as creating better wind conditions for the glider. El Calfate lies just east, or in the lee, of the Andes, and although this is not the highest portion of the Andes, the fact that it is closer to the southern hemisphere polar vortex more than makes up for the relative lack of height. Also, there is no land mass blocking the frontal wind systems that move through the region—like there was in New Zealand—generating the mountain waves. We had chosen New Zealand first because of many factors, such as the support for launching soundings, balloons that helped us forecast the necessary weather conditions, and that the New Zealand National Institutes of Water and Atmospheric Research (NIWA) was located just 40 miles, as the crow flies, away from Omarama. Also, there is a large community of glider pilots and support for glider operations in Omarama. There were other factors, too, such as the availability of heated hangars, the fact that English is the primary language, and the ease of obtaining the necessary equipment and supplies, such as liquid oxygen for the pressure suit rebreather system. In both locations, New Zealand and Argentina, the big issue is that there are only about three days per season (June to October) when the atmospheric conditions will potentially be great, so Argentina's additional odds in our favor for producing these conditions eventually outweighed the other advantages that New Zealand offered.

We had a theory that soaring to altitude into the stratosphere using the polar night jet was possible. Einar Enevoldson had speculated this was the case ever since he saw an image of stratospheric mountain waves, taken from a research aircraft flying in Sweden, hanging on a scientist's wall in Germany. Happily, this theory was proven correct on August 30, 2006, during the long-awaited completion of phase one of the Perlan Project, when pilots Steve Fossett and Einar Enevoldson soared a modified DG505M glider (without a motor) to 50,671 feet out of El Calafate Airport.

"We could have gone higher but the pressure suits expanded so much inside the cabin we couldn't move the flight controls," Einar stated. Einar also told me they were cold, had to go to the bathroom and, most important, had proved their point: that it *is* possible to soar on mountain waves, taking an engineless glider well into the stratosphere.

(Steve confided in me that he did not ever need to use the diapers they wore while flying. The preferred diaper brand of the Perlan pilots: Depends! Astronauts wear them, too, by the way. I agree with Steve: it would be quite difficult to basically pee in one's own pants, so to speak, while also trying to maneuver a glider above 50,000 feet. If there were an emergency, that might be another story. Steve was a friend and had been very enthusiastic about the Perlan Project for many years, so his candor was not surprising.)

Once the pilots decide to come back down to earth, they just move out of the portion of the wave producing lift, point the nose of the aircraft down, and fly back to the airport they departed from. Just as they do when ascending, they must find the best path down, making use of the existing atmospheric conditions at lower altitudes and, hopefully, finding a way to navigate back to the airport from which they departed. In some cases they may be forced to "land out," that is, find a suitable field somewhere to put the glider down safely. This is not an unusual situation for glider pilots.

"They can't find Steve Fossett!" Ed Teets, Jr., blurted out on the phone to me. Ed is the lead meteorologist at NASA Dryden Flight Research Center, recently renamed NASA Armstrong. Ed was a team member on Perlan I.

It had been just over a year since Steve and Einar had broken the altitude record in Argentina.

"What? I don't understand what you mean. They can't find him? What are you talking about?" I asked.

"Well, he went missing in an airplane from Baron Hilton's ranch," Ed responded. I knew all about Hilton's ranch, as Steve had in the past invited me, with my now ex-husband, to join him and his wife, Peggy, there for a weekend. Its official name is the Flying M Ranch. I wish I had taken him up on the offer, as it was supposed to be great fun there, with all sorts of planes, a helicopter, and other great machines, but my ex was not into aviation at all, so we declined. Hilton's ranch in Nevada was not far from our second home on Bridgeport Lake in Bridgeport, California, less than 25 miles away.

It was Tuesday, September 4, 2007, and Steve Fossett had been missing since the previous morning. I remember the day he went missing well. It was Monday, September 3, 2007, the Labor Day holiday. Alan, my husband, Evan, our son, and I were living in Stateline, Lake Tahoe, Nevada at the time. It was the day before Evan was to start his first day at nursery school. He was two and a half years old. That day the three of us rode the Heavenly Ski Resort gondola and hiked up around the top of the mountain. The views of Lake Tahoe were breathtaking. It was beautiful fall day in Tahoe with calm winds, and temperatures in the low 80s Fahrenheit. Three quarters of the way up, the gondola stopped at the observation deck, where we got off. It was around nine-thirty in the morning and we asked another person to take a picture of the three of us. It is one of my favorite photos of our family; I had no idea it was taken at almost the exact moment of Steve Fossett's plane crash to the south of us in the Sierra Nevada mountains. I think of Steve every time I look at this photo.

I still wonder to this day if Steve survived the accident for a time, as the wreckage was not discovered until October of 2008, more than a year after his disappearance. I truly hope he did not suffer. There was quite a bit of confusion as to where exactly he was flying, Nevada or California. When he didn't return by eleven A.M., a search for him began. The whole incident was surrounded by a lot of hoopla, wild stories of him being sighted in South America and other sensationalist tabloid-style absurd speculation as the search dragged on.

After receiving Ed's call about Steve Fossett's disappearance, Einar Enevoldson called me. He asked if I would do a quick reconstruction of the weather in the eastern California and northern Nevada regions the day of Steve's flight and send him a map of possible regions of turbulence to help narrow down the area of their search for Steve. So I did. I looked at model data, weather charts, satellite data, and digital NEXRAD Doppler radar data. There were quite a few turbulent regions over the mountainous areas, of which Nevada has plenty, and couple in neighboring eastern California. One region I circled in particular—a region north of Mammoth Lakes, California—turned out to be where Steve's wreckage was found. But as the search of these highlighted regions began, there were witnesses who made statements that were *mostly* correct, as happens with many accidents, which threw search teams off the trail. One witness saw the aircraft near Yerington, Nevada, which is about eighty miles north of the accident site, but the witness's timing of this sighting was off by one hour. Consequently, this twenty-minute track of the aircraft's flight had been ruled out as a possible site of the accident as a result of this timing error, which could have had serious repercussions if that had indeed been where Steve had crashed.

Steve was flying a Bellanca 8KCAB-180 Super Decathlon, a two-seat aircraft normally flown near sea level and often used for low-altitude aerobatics. About a year after Steve's disappearance, the accident site was discovered by some hikers near the Minaret Range just northwest of Mammoth Lakes, California. Adding to my grief was the fact that I knew this area of the country well. It is composed of steep, rocky mountains and cliffs dotted with beautiful lakes. The mountains here rise up to over 13,000 feet in elevation. Around the time of the accident, the meteorological factual report from U.S. National Transportation Safety Board (NTSB) concluded that the aircraft encountered downdrafts of at least 400 feet per minute and calculated that at the altitude the aircraft was flying (10,000 to 13,000 feet), it could only handle—at most—downdrafts of 300 feet per minute. The force of the downdrafts was too strong at these altitudes; Steve's small plane crashed into the hillside at a high speed.

The NTSB's official probable cause of the accident: "The pilot's inadvertent encounter with downdrafts that exceeded the climb capability of the airplane. Contributing to the accident were the downdrafts, high density altitude, and mountainous terrain." I heard later that I had supposedly tried to contact the NTSB to find out about the accident, to state that I agreed with these conclusions. I couldn't believe my ears. I had never tried to contact them, but I knew immediately who this impersonator was, pretending to be me. It's something I seldom mention but still think about. This left me feeling betrayed and, frankly, angry that someone would stoop to these levels just to get information about the accident. I am over it now and just feel sorry for this always-to-remain-unnamed person.

It took nearly seven years after Steve Fossett's death until the Perlan Project was officially back on track. It took Dennis Tito as an investor and pilot, along with retired United States Air Force test pilot Jim Payne acting as chief pilot, to breathe life back into the project. You may remember that Dennis Tito was the first American tourist to go into space. In 2001, he went with the Russians on a Soyuz spacecraft and spent about seven days on the International Space Station before coming back to Earth. In 2014, when Airbus Group became Perlan II's title sponsor, Dennis became interested in providing the final investments, and we were off to the races. Airbus's sponsorship has enabled completion of a new carbon fiber experimental aircraft. The Perlan I glider was a modified off-the-shelf glider. The Perlan II has been completely designed and built from scratch for its mission to get to 90,000 feet. The world absolute altitude record attempts begin in El Calafate, Argentina, in July of 2016, nearly ten years after we set the altitude record in the same spot. This new aircraft is designed to fly efficiently at low attitudes and at very high altitudes. It has a wingspan of 84 feet and will have a cabin pressure of 8.5 pounds per square inch (an atmospheric equivalent of 14,000 feet). The aircraft will have a drogue chute that can be deployed in case of an emergency, which will bring the aircraft down safely to the ground. Unlike in the Perlan I glider, the two pilots will not wear parachutes, as they cannot escape from the fuselage, but rather would have to come down with the ship. This design feature is in place so that the aircraft can be pressurized. The aircraft is designed to fly at peak performance at approximately 60,000 feet, as this is the region where we expect the coldest temperatures (as low as -90 degrees F) and also the possibility of the strongest turbulence. The glider's maneuverability to avoid this turbulence will be paramount.

The breaking of mountain waves causes extreme turbulence, which the Perlan II will be trying to avoid. Just as waves in the ocean crest and break, the mountain waves travel higher and higher into the atmosphere and become steeper and steeper, eventually breaking and crashing down. Breaking waves are a known source of tremendous turbulence, even on commercial flights, when the mountain waves break down lower in the atmosphere. But the higher the waves travel, the more likely they are to break, putting the Perlan II glider at risk as it maneuvers up to 90,000 feet. It will be my job, and the job of other atmospheric scientists at WeatherExtreme Ltd., to provide accurate and timely forecasts and information to the Perlan II pilots so as to avoid breaking waves that could put the mission, as well as the pilots' lives, at risk.

Forecasting for turbulence is not straightforward or exact, especially at these altitudes and in these remote locations. First, there are very few actual measurements other than via the weather balloons that are launched twice daily, but neither balloon is very close by. One balloon is launched in Argentina, 370 miles northeast of El Calafate, offering not much help, and the other one is launched in Chile, well over 600 miles north of El Calafate, which is of even less service to our cause. The next nearest weather balloons are launched on the Antarctic continent. Hence the importance of our Perlan team launching our own balloons, preferably in Chile, upwind of our location. It has to be upwind so that when the balloon is released it travels through our intended flight airspace and can give us the most exact readings possible. Not only do we then have an actual measurement of the vertical profile of the atmosphere in which we will fly, but we will also have data that we can feed into the high-resolution mesoscale meteorological model, WRF (Weather Research and Forecasting System), that we will be running in near–real time. This will help us forecast upcoming wave days as well as help forecasters keep the aircraft in the locations of best lift and away from regions of potential breaking waves.

I first came to appreciate the difficulty of identifying atmospheric turbulence from one of my professors in graduate school, Dr. James W. Telford (1927–1999). Back in the 1960s Dr. Telford and Dr. Donald H. Lenschow pioneered the technique for catching turbulence in action through aircraft-mounted probes. Dr. Telford spent much of his career trying to understand and dissect the nature of atmospheric turbulence. I think Dr. Telford would get a chuckle out of the apocryphal story in which Werner Heisenberg, the German theoretical physicist, was asked what question he would ask God, given the opportunity. His reply was, "When I meet God, I am going to ask him two questions: Why relativity? And why turbulence? I really believe he will have an answer to the first." Turbulence still holds many mysteries, even with the help of modern technology, but should we succeed in the Perlan II's mission, we will have yet another opportunity to learn more about the upper reaches of our atmosphere.

If and when we fly our Perlan II at our goal of 90,000 feet we will set a new world altitude record for manned, non-astronomical flight, currently held by an SR-71 Blackbird, a long-range, Mach 3.2–plus jet powered aircraft, of 85,069 feet set back in 1976. Even as we pass through the 60,000- to 70,000-foot region of the atmosphere—the most turbulent—the danger does not end there. As the glider nears 90,000 feet it will be dangerously close to reaching its "coffin corner." The two limiting parameters of what aerodynamicists refer to as the "coffin corner" are stall speed and Mach limit. Stall speed is a condition in aerodynamics in which the angle of attack of an airfoil (like a wing, for example) increases beyond a certain point, at which the lift begins to decrease and the airfoil is no longer flying—it stalls. Mach limit is the limiting speed of a particular airfoil, beyond which it will potentially come apart. The aerodynamic characteristics of an airfoil change radically and unpredictably when the airfoil exceeds the limiting speed for which it was designed. In order to maintain lift at these extreme altitudes the airframe of the Perlan glider will be approaching transonic speeds. Basically, the glider will be on the edge of stalling if it goes any slower and on the edge of going Mach and potentially breaking apart if it goes any faster. Hence, the ominous name "coffin corner." As the name implies, this is not a region where one wants to venture.

But despite the tremendous risk and complex confluence of conditions necessary for such a glider flight, the atmospheric data that will be collected in the process will be extremely valuable and can then be analyzed to help unravel the many mysteries still eluding us as to the particulars of the ozone hole, pollution, and climate change. We will address those mysteries by running experiments to gather data about heat, mass, and chemical exchange between the troposphere and stratosphere that can be used to improve our climate models. I agree with Robert Ballard, professor of oceanography, who is quoted at the beginning of this chapter: the age of exploration on our planet is just beginning!

TWO

The Business of Weather

Some people are weatherwise, but most are otherwise.

—Benjamin Franklin

It was 1994 and I had known for years that my ultimate goal was to start my own weather company, something I had been thinking about since I started college. Both my parents went to Duke University—it's where they met. It was one of the handful of universities to which I had been accepted, and I knew they would have loved it if I followed in their steps. It was a difficult decision, but I went with my heart and chose the University of California system and wound up at UCLA. Over the years I've never had to second-guess that decision because it's clear to me that I made the right one. It was during my undergraduate studies there I met many giants in the field of weather, and my training got me off to a wonderful start and solidified my belief that I had made the perfect career choice.

By 1993, I had actually completed all the work toward my PhD at the University of Nevada at Reno, but not all of my advisors were able to sign the final accepted version of my dissertation because two of them were on research trips out of the country. Finally, the absent two were able to sign and I was then officially Dr. Elizabeth Austin. It was done! Walking into the graduate school office at the University of Nevada, Reno to turn in my fully signed and completed PhD dissertation was like walking on air. What a relief to have finished all of my courses, successfully taken the comprehensive exam, and then composing the final note in this symphony of study: the signed dissertation. I would finally receive my doctorate in atmospheric physics. (Though the degree was through the physics department and on paper read "physics," my degree was actually "atmospheric physics," because at the time the university did not have a separate atmospheric science department.)

I graduated that summer with all my family in attendance. At the ceremony my brother Patrick heard a woman standing next to him say, "Look at that girl—she's getting a doctorate in physics!" Patrick proudly said, "Yeah, that's my sister." Patrick, my lifelong friend and ally, was soon to become a very important part of the company that was gestating in me.

I received the actual "onionskin" diploma the following year when I was already well into my postdoctoral work at the Desert Research Institute's (DRI) Atmospheric Sciences Center in Reno. The dream to start my own weather company became stronger and stronger and, in September of 1994, after a year of postdoctoral work at DRI, I made my dream a reality and founded my company. I named it Firnspiegel, a name, perhaps, that only a scientist could love and one that would probably make a marketing director cringe. But it made sense to me at the time. *Firnspiegel* essentially means "mirror snow" in German and is a common type of snow formation in the springtime. Firn is an older type of snow and speigel means mirror in German. Firnspiegel is a thin sheet of ice on the surface of the snow caused on sunny, cold spring days as the sun's heat penetrates the surface snow but then meltwater at the surface is refrozen, forming a thin layer of ice. This requires just the right heat balance and creates a shiny snow cover that looks like a mirror when the sun hits it just right. At the outset, Firnspiegel was just me, but I founded it knowing that I wanted to perform atmospheric research and also provide weather forecasting and forensic meteorology services.

I left DRI and took a three-quarter time teaching position at Sierra Nevada College (SNC). The private four-year college is located on the beautiful north shore of Lake Tahoe in Incline Village, Nevada. Not only did the teaching position provide some needed income but, because the college encouraged its professors to do research and, more important for me, consulting outside the college, I was allowed to start gaining some "real-world" experience and spend time developing Firnspiegel. I had a vision to create a company that provided a variety of weather-related services in the areas of forensics, forecasting, and other problem solving areas tailored to specific client needs.

Through teaching at the college I was able to build my company so that a few years later I turned it from a sole-proprietorship to a limited liability company (LLC). I actually now had employees at Firnspiegel LLC. It was no longer, me, myself, and I running the show. I now had six people who would help me with our various weather projects, including forensic meteorological services, weather forecasting for specialty projects, and problem-solving for companies who were struggling with issues relating to weather, such as outages in satellite signals or choosing the appropriate location for their headquarters. It was a very exciting time for me. With my parents sharing in the initial real estate investment, I bought a commercial building on the north shore of Lake Tahoe and was off to the races. Though I had that support for the building, the company itself has never had any financial assistance: it was formed and run on my "sweat equity" alone. I had found my *raison d'être*—my reason for living. Now that I have a family, today my reason for being is family and then career. I have to confess, however, that I have brought my immediate family into my career, a merger that makes me very happy.

Even though I was still fond of the name Firnspiegel, as time went on I became tired of having to pronounce and spell it to almost everyone—no problem if we were doing most of our business in Germany, but we weren't. My husband, Alan, also had some thoughts and humorous takes on the difficulty of using the name Firnspiegel. When I told him about some of the other names I had reserved, we both agreed that one of them was perfect: WeatherExtreme. So in the early 2000s, we trademarked and changed the name to WeatherExtreme. WeatherExtreme Ltd. grew dramatically in expertise and services, and now offers weather forecasting, forensic meteorology, and weather research services, but its corporate goals have remained the same since its inception. Alan never lets it go unsaid, however, that the company is still "me." It is the guidance, expertise, and credibility of Dr. Elizabeth Austin that is in demand—WeatherExtreme Ltd. is the wrapper that the candy is wrapped in. That, however, I am happy to say is changing as the company moves forward and gains talented and qualified employees who bring further expertise and insight into the fold. Being the only expert was in the beginning quite flattering, but as business has increased dramatically, it's also quite limiting and burdensome. In order for the company to grow, our new employees would have to be the critical element for the future of WeatherExtreme Ltd.

It is perhaps a lucky coincidence that more than three decades ago I chose the field I did. Weather has since become a hot topic outside the scientific community. It is truly amazing how governments, industry—all of us—depend on weather and climate. It is even more startling how few people are aware of how much they are affected by weather. The effects are global and include health, well-being, and, of course, tremendous financial implications. This is where the business of weather comes into play. Weather is a business and it is *big* business—which is one of the reasons it is often the lead story on the morning and evening news.

Weather and climate used to be solely the province of governments, when they were the only entities to provide official weather forecasts for the general public and the military. The military is where most meteorologists used to be trained, but in today's world, private weather companies are far surpassing the government in terms of dollars, as well as the number of people (and companies) turning to them to understand weather and how it affects their lives and businesses. Global economies are directly and indirectly affected by weather and climate change, whether it is extreme weather events or just the seasonal events, such as winter storms and their impact on travel, tourism, and logistics (including the shipping of packages). The World Bank has reported that extreme weather events alone cost the global economy two hundred billion dollars per year. Back in 1998, in testimony to the United States Congress, former commerce secretary William Daley stated, "Weather is not just an environmental issue; it is a major economic factor. At least $1 trillion of our economy is weather sensitive." Dr. J. Marshall Shepard, past president of the American Meteorological Society, worded it slightly differently: "About one third of industry and business today is weather-climate sensitive." The economic, as well as national security, impact of weather is now at the forefront of national discussions involving many sectors of the U.S. government, including NASA, the military, and the Department of Commerce.

Weather is such big business that if one thinks of it like stocks or futures, then very small hedges can produce huge returns. For example, if Home Depot has knowledge that the northeastern United States will most likely experience a snowier than average winter, then they can make a number of cost-saving and, more important, money-making decisions. They can ensure that all of their stores in the northeast have plenty of snow shovels, heaters, and winter items. But they can also be ready for the onslaught of winter by having the stores stocked early in the season and not playing catch-up all winter long.

Often the implications of a particular weather forecast may not be so commercially obvious. For example, there are software programs that use the air temperature along with past crime locations, such as muggings, car thefts, and robberies to predict areas where crimes are likely to occur or increase. When the weather heats up, crimes increase.

Industry, corporations, and even individuals can capitalize on accurate weather forecasts that describe changes of even just one degree warmer or colder! There is a saying in meteorology called the "profit of one degree." It sounds trite, but this theory is based on research and facts. Some are seemingly bizarre but true. Here is a portion of a list, compiled by WxTrends.com (note: "wx" is shorthand for "weather"), that shows what happens when the temperature *warms* by just one degree in developed countries:

- More than 10 percent more fans sold each week
- More than 11 percent more sun care products sold each week
- More than 4 percent more fresh strawberries sold each week in summer

- 22,424 more automotive batteries fail each week
- More than 6 percent increase in fire ant products sold each week
- More than 2 percent increase in energy consumption (electrical power used) each week

If the temperature *cools* by one degree:

- More than 3 percent more orange juice sold each week
- More than 25 percent more mousetraps sold each week
- Over 15 percent increase in portable heaters sold each week
- 5,000 more units of medicated lip care products sold each week
- 1,000 more vaporizers sold each week
- Over 2.5 percent increase in total company soft line (merchandise such as clothing, footwear, jewelry, linens and towels) sales each week

Whether they all know it or not, the fate of retailers, whether they sell cosmetics or pesticides, is intimately tied to the weather and climate. If it is too cold, shoppers tend to stay home (although now with online shopping becoming more popular, these statistics may change). Weather is tied to retailers' profits, margins, demands, inventory, planning, and expansion—nearly every aspect of business. The benefits of having an accurate weather forecast: better inventory management; better decision making on purchases, distribution, marketing, and planning; the ability to sell at higher margins; the ability to identify new markets, opportunities, and products; and cost savings *and* higher profits.

But even if the value of climate prediction and how it relates to the upcoming season is not so immediately apparent, merchants and others can still take full advantage of meteorologists' intimate knowledge of weather by obtaining accurate predictions in a timely manner. Sounds simple, but it takes planning and persistence. This planning involves being in constant contact with a weather center as potentially severe weather is forecast. Why? So that, for example, a company like Lowe's can have an advantage over Home Depot by making last-minute changes to its schedule and routing products to various locations ahead its competitor. If there is a cold dry air mass moving into the midwestern United States from Canada, and a warm, moist air mass flowing up from the Gulf of Mexico via the Gulf Coast, states where these two air masses merge can expect severe weather. The trick is to accurately forecast where and when this air mass battle—and the severe weather (including tornadoes)—will occur. The products that people require during severe weather are numerous and vary greatly, from the basics of water, toilet paper, and emergency radios to plywood, sandbags, and generators.

Overall business trends improve when the weather gets better, yet employers of all business types should relish when the weather gets *bad.* It has been proven that when the weather is bad, that is, rainy, cold, or snowy, workers become more productive. It makes sense: there is less to distract them. In an interesting corollary finding, Dr. Jooa Julia Lee of Harvard University conducted research on how productivity is affected by the weather. She found that even if the weather outside was rainy and gloomy, productivity would increase, but if workers were shown photographs of outdoor activities and fun in the sunshine and were asked to imagine themselves doing those activities, their productivity declined! Thus, weather—even just imagined weather—and its effect on people is the ultimate mind game. Similarly, there have been many studies relating human productivity, attitude, and habits to where they live in relation to the equator. Hot climes produce vastly different effects from cold ones on humans.

Many studies have found a direct correlation between hot temperatures and human aggression. In fact, Dr. Solomon Hsiang of Princeton University and his colleagues from the University of California Berkeley discovered that there is a universal relationship between rising temperatures and increased human conflict and social unrest. These results indicate the huge implications of the climate change issues we are facing. Interestingly, there is also evidence that extreme rainfall events also cause an increase in human conflict. So if you have a hot clime with periods of extreme rainfall events, look out.

Cold climes, on the other hand, may affect your health. People tend to be more sedentary in colder climates, so that when they do exert themselves, there is greater risk of heart attack, as they are not conditioned for physical stress. Compounding this risk is the fact that your heart has to work harder in the cold to help maintain your body's core temperature. Chilly temperatures also cause the blood to thicken and arteries to constrict, elevating one's blood pressure. However, people who live in cold climes take note: countries farther from the equator, in colder climes, are wealthier than countries near the equator.

There is the obvious impact of weather on the agriculture, energy, and insurance sectors but there are many other less obvious impacts of weather on other businesses. For example, the intersection of weather and sports—talk about big business! The weather can have an effect on the score of a football game, the number of rounds of golf played in a season, and the number of skis sold, not to mention the effect it has on ski resorts. Weather is now even a commodity on the Chicago Mercantile Exchange and on European stock exchanges, which trade in various products such as monthly and seasonal futures based on temperatures!

To take things a step further into the world of big business, corporations, and huge conglomerates that control many of the items we rely upon, let's think about this: what if a handful of these companies could gain an advantage over the National Weather Service or other large, privately held companies regarding weather forecasts? How could they do that? Something as commonplace as a watch could provide the key. If everyone had a watch that monitored air temperature and humidity in the wearer's location, and these watches were worn by millions of people, and then the data from these watches was fed into a central computer and aggregated, these data could then be used to monitor the weather conditions at the surface practically in real time. It could also be fed into weather forecasting models for more accurate and timely forecasts, which the company controlling this central database could then take advantage of for maximum profit. Many companies are already taking steps toward this end, such as beginning to utilize the tracking devices used by trucking companies and county and state vehicles, including snowplows, to monitor the weather and road conditions in near real time.

Automobiles are another mass-market item that can be used to monitor weather data. Many cars are now equipped with black boxes, similar to the ones found in aircraft, which monitor many functions, including your location, the time of day, and your motion (speeding, braking, etc., as measured by an accelerometer). It would be fairly simple to add the measuring of weather parameters to the mix, including how the tires of the vehicle sense the road in different conditions. This is already being done with sensors on test and research vehicles. These are great technological advances that also have some potentially scary consequences, like monitoring one's every move when in one's vehicle.

Many cell phones now come with barometers in them. These tools are great fun for the individual but are potentially extremely profitable for the owners of the data they generate (which isn't necessarily the cell phone owner). Weather companies are creating apps that measure the weather conditions where one is located and they are archiving these data. Weather data is so voluminous these days that it falls under a term now used for data aggregations of this type—"big data"—which covers everything from website search histories to the business of weather. Archiving actual weather data alongside high-resolution atmospheric models, raw weather data collected from balloons, ships, aircraft, and even buoys in the ocean—and running climate models covering the entire planet, which generate terabytes of data in just a few minutes—is a business in and of itself. The masters at big data are companies like Google, Amazon, and Yahoo! Whether we like it or not, these companies and others like them are poised to become larger and even more powerful, using the weather as a tool, if they choose to do so.

Aside from the commercial implications of weather data collection, the weather business also involves disaster planning. According to an Ad Council survey, almost two thirds of small businesses do not have an emergency plan in place in case of a disaster. Flooding is the most common disaster in the United States and causes approximately fifty billion dollars in annual economic losses each year. Another study by the Insurance Information Institute found that 40 percent of businesses affected by manmade or natural disasters never reopen. The U.S. National Oceanic and Atmospheric Administration (NOAA) has a new initiative called "Weather-Ready Nation." WeatherExtreme Ltd. is a Weather-Ready Nation Ambassador as part of this initiative, which is a nationwide effort to formally recognize NOAA partners who are improving the nation's readiness against extreme weather, water, and climate events. As a Weather-Ready Nation Ambassador, we are committed to working with NOAA and other ambassadors to strengthen national resilience against extreme weather. In fact, it does not take a major disaster to cause problems. Emergencies can be created by a couple of moderate, or even minor, events coming on the heels of each other. For example, if there is a moderate earthquake that isolates certain communities through road collapses, and then a windstorm follows, creating wildfires . . . now there is a disaster brewing. It is critical to build community resilience in the face of increasing vulnerability to extreme weather events, including flooding. One way of building community resilience is improving communication between the National Weather Service and stakeholders and the general public. Another is developing advances in meteorology for greater accuracy in the forecasts and weather warnings issued. Other ways include teaching in schools and universities about the risks of severe weather and how to be prepared. Companies such as WeatherExtreme Ltd. are working with community leaders to develop emergency preparedness plans and with insurance companies providing discount incentives to policy holders who meet certain mitigation criteria.

Why not just consult a Farmer's Almanac? Actually there is the *Farmer's Almanac* and there is the *Old Farmer's Almanac.* They are different. Since 1792, *The Old Farmer's Almanac* has published yearly, and is issued on the second Tuesday in September. Over the years there have been many competitors. Probably the most popular is *The Farmer's Almanac*, which is an annual publication for North America and has been published since 1818. Both publications contain a lot of predictions, whether for colder and drier winters in the Midwest or a wetter, warmer winter in the southwestern United States. Both publications include predictions for weather; the *Old Farmer's Almanac* is famous for its long-term forecasts. However, what can be said about both publications is that they are more likely best used for climatological-type forecasts—for example, will New York City be warmer than normal this winter—rather than meteorological forecasts—is it going to snow in Chicago on March 15th? But of course more sophisticated data analysis is now available, especially with the advent in the twentieth century of supercomputers that can run highly sophisticated, high-resolution weather models. Both Farmer's Almanac publications claim 80 percent accuracy, but neither actually publishes these results; they only give these "statistics" out in interviews. Numerous scientists have studied the accuracy of these publications and they have shown that both almanacs' predictions are no more accurate than those generated by a game of chance. Nevertheless, I still enjoy reading these prognosticator publications. It is kind of like reading a weather horoscope. Considering that these publications are only produced yearly, it is now impossible for them to truly compete or provide accuracy in today's big data market.

So how does the charming weather lore of the past compare to the skill of weather forecasters? Take Punxsutawney Phil, the furry friend who shows up annually each Groundhog Day to let us know if we should expect six more weeks of winter or not. The U.S. National Climatic Data Center, which archives much of the weather and climate data for the country, analyzed the actual data versus Phil's predictions. They concluded that this forecasting groundhog "has shown no talent for predicting the arrival of spring, especially in recent years" and that "Phil's competitor groundhogs across the nation fared no better."

But not all weather lore is to be ignored. Many stories and proverbs have turned out to be accurate observations of weather and climate. For example, "Red sky at night, sailors delight; red sky in morning, sailors warning." This rhyme is based on the fact that sailors needed to watch and understand the sky when predicting the weather and thus sailing conditions for the next day. In mid-latitudes, the weather generally flows from west to east, so when a sailor sees a red sky at night, when the sun is setting, it is a visual cue of high atmospheric pressure and thus a harbinger of generally favorable weather coming in from the west. To get this red sky, there must be stable air, a sky that is mostly cloud free, with the high concentration of dust in the air creating this red sky. But if there is a red sky in the morning, when the sun is rising in the east, this means that the high pressure and good weather is east of the sailor, not west. This means that a low pressure system is moving in to the sailor's location, pushing the high pressure out of the way and bringing bad weather with it. However, this does not hold true for lower latitudes, where the weather patterns move from east to west.

—

Running one's own weather company, or any company for that matter, is not for everyone. It means long hours, usually no retirement or 401k, and having to be a jack-of-all-trades. In the case of my company in particular, I've found the need to be scientifically weather-savvy as well as business- and marketing-savvy in order to compete. Especially in the beginning, running a business is not glamorous; it entails being the receptionist, the administrator, the consultant, the expert, and even the janitor! But I wouldn't trade it for the world. Today WeatherExtreme Ltd. works on a variety of projects and cases, such as forecasting for world record events, solving weather-related problems for agencies such as NASA, and reconstructing the weather surrounding murders, plane crashes, and various extreme weather events. We are in the business of weather, and it's everywhere! It's not just a small-talk cliché. One cannot escape it. Our daily lives, how we feel, how productive we are, and our sheer existence depend on it. The three biggies are: weather, water, and climate. These three things are going to affect us and our planet in the years to come in a way society never thought possible.

I didn't aspire to become an atmospheric physicist because I thought weather was a "good business." I got into the field because I have a scientific bent and loved the study of the atmosphere. It turned out to be a wonderful decision both intellectually and practically, because not only do I love it, and have been able to make a career out of it, but it has also allowed me to try to make the planet a better place to live in, at a time when understanding weather has never been more critically important to the health of our society.

THREE

Where's the Dead Body!

In a well-balanced, reasoning mind there is no such thing as an intuition—an inspired guess! You can guess, of course—and a guess is either right or wrong. If it is right you call it intuition. If it is wrong you usually do not speak of it again. But what is often called intuition is really an impression based on logical deduction or experience. When an expert feels that there is something wrong about a picture or a piece of furniture or the signature on a cheque, he is really basing that feeling on a host of small signs and details. He has no need to go into them minutely—his experience obviates that—the net result is the definite impression that something is wrong. But it is not a guess, it is an impression based on experience.

—M. Hercule Poirot, *The A.B.C. Murders*,
AGATHA CHRISTIE (1936)

"Make me look as old and trustworthy as you can," I told the photographer. He looked at me, flabbergasted, and replied, "In all the years that I have been taking pictures of women, I have *never* been asked by a female to make them look *older*." I was at the beginning of my career in forensic meteorology and I needed the right kind of head shot for my curriculum vitae—my detailed résumé.

I had no idea then that the breadth of cases I would be asked to work on over the course of my career—more than 1,500 to date—would include dealing with some sort of calamity, whether it was major airline crashes, boating accidents, kite surfing deaths, midair plane crashes, tornado and straight-line wind deaths and destruction, huge automobile pileups, building collapses, attempted bombing of an Internal Revenue Service building, and even a double murder, which would result in the death penalty. Since venturing down the path of forensic meteorology, I've gotten a variety of interesting questions posed by a diverse array people, far outside the realm of what one would think of in terms of meteorology, or even the sciences.

"Where's the dead body?" a man named Jim Ashby asked me in 1994 as I walked into the Western Regional Climate Center at the Desert Research Institute where Jim and I then worked as colleagues. This comment reminds me of the case that really launched me into the area of forensics: the crash of Korean AirFlight 801. On August 6, 1997, over Guam, a Boeing 747-300 aircraft crashed into the side of a mountain while attempting a landing during a storm, killing 228 people; only 26 souls survived. I was hired by the plaintiffs' (victims and families of the victims) steering committee to reconstruct the weather conditions surrounding the flight and accident.

Guam is an unincorporated territory of the United States, located in the western Pacific Ocean and part of the Mariana Islands. Much of the island is surrounded by coral reefs. The island is thirty miles long and ranges from four to twelve miles wide. Guam is more than 5,800 miles west-southwest of San Francisco, California, and is 1,550 miles south of Tokyo, Japan. The highest point on Guam is Mount Lamlam, at 1,334 feet elevation. There are brown tree snakes, known as *kulebla*, all over the island, which most likely arrived as stowaways aboard U.S. military ships or aircraft shortly after World War II. These snakes average between three and six feet long and have taken over the island. Though the snakes are harmless to humans, they have killed off most of the native birds and have become a pest to the island's ecosystem. For a time the government of Guam was offering monetary rewards for all brown tree snakes captured and turned in to officials. But that plan backfired as the locals then began raising these snakes and killing them in order to cash in. It is said that there are more than twenty snakes per every acre of land on Guam. The island is 212 square miles, so this calculates to more than 2.7 million snakes on this small island! The newest tactic to fight the snakes is to drop dead newborn mice from helicopters. The mice are implanted with acetaminophen (yes, the drug in Tylenol), which kills the snakes if they ingest it. The mice are outfitted with tiny parachutes made of green tissue paper and cardboard (an interesting image worthy of a social media video) and dropped into the jungle below. Their parachutes are designed to get caught up on the trees of the jungle canopy where the mice become dinner for the unwitting snakes. This has apparently helped reduce the snake population slightly, but it will probably not yield a win in the battle against this snake explosion there.

After learning of the Korean Air crash, my team and I began to piece together the airline's timeline. Just after 1:42 A.M. local time, Korean Air Flight 801 was on approach to the A.B. Won Pat International Airport in Agana, Guam, when it crashed into Nimitz Hill about three miles southwest of the airport. The plane had been cleared for landing on Runway 6 Left and was on final approach when it hit the terrain and exploded in a ball of flames. The plane was completely destroyed on impact. The elevation of the destination airport is 297 feet above sea level and the elevation of the accident site was 660 feet above mean sea level. The aircraft had departed the Kimpo International Airport in Seoul, Korea, at 8:53 P.M. (9:53 P.M. Guam time). During the final portion of the flight, the aircraft was flying through heavy rain. A certified navy observer on Nimitz Hill was about three quarters of a mile from the accident site and stated that at the time of the accident the cloud ceilings were about 700 to 800 feet, or around 160 to 260 feet above the terrain where the plane struck the ground. (Cloud ceilings are measurements of cloud bases relative to the ground.)

The navy observer also stated that the visibility was around 200 to 300 meters, or about one tenth to two tenths of a mile, and that the winds were not more than 10 knots. The aircraft was in Instrument Meteorological Conditions (IMC), meaning "in cloud," during this final portion of the flight. There was also another rain shower in front of the plane, between the plane and the airport, affecting the pilot's ability to see the airport. Normally this should not be a problem for an experienced pilot, but the captain was very tired, which was a likely component in his poor performance. He was expecting a visual approach to the airport; the pilots did not expect to have to use the instrument approach, so they weren't fully prepared to fly one. There were also other issues with instruments and the air traffic controller on the ground in Guam. The glideslope portion of the instrument approach was out of service, but the captain believed it was in service and picked up a stray electronic signal from an unrelated device on the ground, which he was following. Other members of the crew tried to alert the captain that something was amiss, because the airplane was descending very rapidly—not a typical profile on an instrument approach. The captain ignored their observations and pressed on until impact. As is often the case, this accident was a deadly combination of weather events, pilot error, incomplete pilot training, and air traffic control issues. The captain ignored one of the most important things in commercial aircraft flying: crew coordination, also known as crew resource management. He didn't listen to his crew. That was one deadly element in the crash. Another was a ground/air traffic control issue: the Minimum Safe Altitude Warning system (MSAW) at the airport had been modified to limit spurious alarms and, as a result, could not detect aircraft that were below minimum safe altitudes on the approach. As is the case with all crashes, it was a combination of elements: the crew's poor execution of a non-precision approach in very poor weather, using outdated navigational maps, and poor ATC (Air Traffic Control) monitoring of the flight path of the aircraft.

In a case like the one involving Korean Air, I appear as an expert witness and am deposed by the opposing side. All named parties in a lawsuit have the right to conduct discovery, a formal investigation, to find out more about the case; this is the purpose of a deposition. Depositions, during which sworn evidence is given and the opposing side's attorneys ask questions of the witness or expert to prepare for the big day in court, are a fact of life for a forensic meteorologist, or any expert witness, and can be very stressful and contentious. Depositions of experts are not always taken, for many reasons, in which case the expert just testifies at trial without giving a deposition. This can be for cost savings, or because the opposing side does not feel the expert's testimony is a threat, or believes that their equivalent expert will be more persuasive.

Unfortunately, some attorneys can become quite aggressive during depositions—more so than they might be in front of a jury. Juries can be rather unaccommodating to attorneys who "beat up" witnesses—especially sympathetic female ones. I was in the midst of having my deposition taken for the Korean Air case when I suddenly started getting a terrible sore throat, and I was finding it difficult to speak—bad timing. I forged on and, to make matters worse, the deposition took all day. I had my laptop and projector and was showing the various weather graphics and animations that I had put together for the deposition. (This deposition took place before most attorneys had projectors in their offices; consequently I lugged mine around everywhere. But it paid off, as I made some great connections with law firms and lawyers with whom I would work as an expert witness on many future cases.)

Not being an attorney (perhaps thankfully so), and thus not initially very familiar with the legal procedures, I required some considerable indoctrination to become not just an expert but a savvy expert. From the beginning I was able to maintain a certain demeanor and equilibrium on the stand while sparring with attorneys and not letting myself be bullied or overwhelmed. It was my very first trial testimony, and it was a bench trial (meaning no jury, just a judge). There was one other expert, Dr. Lindley "Lin" Manning, testifying as well. He is an expert in accident reconstruction who at that time had testified in courts all over the United States and was a seasoned veteran of the courtroom. After I completed my testimony, including the cross-examination, and stepped off the witness stand to join him in the gallery seats, he leaned over and said to me, "You are going to have a very long and successful career as an expert witness." The stakes are very high in these cases. I have always been keenly aware of the necessity of maintaining a balance between defending the work I've spent months or years preparing and coming across as a sympathetic human being at the same time, as these cases often involve a loss of life.

Prior to the Korean Air class action suit, I had previously testified as a weather expert at a trial in Ely, Nevada, in front of a judge, in lieu of doing so in front of a jury. Luckily, I felt quite comfortable my first time out of the gate. That initial experience got me started off on the right foot, and I began to learn some of the tricks and turns of phrases that attorneys use when expert witnesses are on the stand. Some of them are so overused by lawyers they must have been taught in Lawyer 101. For example, the phrase "as you sit here today . . ." is used constantly. In the twenty-plus years since my first trial testimony, I have testified in so many depositions and trials that when I hear an attorney preface his or her question with this phrase I cannot help but laugh to myself. Another trick they often use is to call me "Ms." or "Mrs." when they know quite well that I am Dr. Austin. That forces me into a decision: do I politely remind them of my doctorate and risk looking pompous in front of the jury or do I let it go, and risk the jury forgetting my professional credential?

Most of the time, however, the tactics that attorneys use are more subtle. Once, I was testifying in a case in Richmond, Virginia, about a plane crash. I had been retained by the defense, whose firm represented Honeywell International Inc., the manufacturer of navigational and control components for the aircraft. Toward the end of my testimony, as I was being cross-examined by one of the plaintiff's attorneys and he was winding down his questions, he asked me, "So, Dr. Austin, are those all of your thoughts regarding this case?" I heard it immediately. *Thoughts.* I calmly answered, "Those are not my *thoughts.* Those are the *facts* of the case." After I returned home and the trial ended, I was informed that the jury came back after less than an hour of deliberation with a verdict in favor of Honeywell.

Over the years I have worked on a remarkably wide variety of cases, but there are some that haunt me. They are the ones with file folders containing pictures of the casualties from these accidents. In a graphic and terrible way, these photos make all too personal the tragedies that are part of the world of forensic meteorology. The worst ones are roadway accidents and aviation accidents. Though most planes do not just fall out of the sky while cruising in flight, sometimes it does happen. In-flight breakups are rare but they do occur, and I have worked a number of them. The photographs of the scene are necessary viewing, as they provide clues to the weather conditions around the time of the accident. Then it becomes my job to analyze whether the weather was the cause of the crash or if there were other factors at play, be they mechanical failure or human error (or worse—intentional acts). The photographs of bodies from these types of accidents are the worst, and still stick in my mind.

Photographs of the scene of a plane crash may be used to determine who was in which seat of the airplane. This is sometimes a point of contention as, at times, it is not known who was actually flying the aircraft. In addition, in some cases, the injuries to the bodies are analyzed in great detail in order to determine what is deemed in legalese as "pain and suffering." It becomes necessary for some accident and medical reconstructionists to go through the injury list and photographs of the bodies in extreme detail for the jury to determine how long victims lived after the accident and, more important, whether they were conscious and suffering during this time period. If they were awake and yet had mortal injuries, they might suffer for a matter of minutes (the time it takes for one to die if, say, they cannot breathe), or dying may take hours or days. I have worked some horrendous cases in which it was obvious that some of the victims suffered for days, but a rescue could not take place due to inclement weather or difficult terrain. By the time rescuers arrived at the scene of the accident it was too late. Every now and then when I am on a flight I think of these images and immediately try to put them out of my mind.

Helicopter crashes can also be especially grisly. To make it even more emotionally fraught for me, my husband, Alan, is a professional airplane and helicopter pilot. He spent many years flying for private charters as well as on movie sets, where he flew, acted, and did aerial stunts in several movies and TV commercials. Alan enjoys flying helicopters, but he describes flying them as "like trying to fly a washing machine during spin cycle": the feeling is that all the parts are spinning around in a concerted effort to destroy themselves. The majority of the helicopter crashes that I am called on to analyze occur either in mountainous terrain (including Hawaii) or are due to the helicopter hitting electrical high-tension wires. Hawaii has more than its fair share of helicopter crashes because there are so many helicopter scenic tour operators, along with the fact that the weather there weaves its way quickly through some very challenging terrain.

Everyone has probably seen those large, bright orange balls that are strung along high-tension power lines in canyons and valleys. These wires pose tremendous hazards to helicopters and result in many accidents. Even Alan has had near run-ins with these wires. One cloudy night, he was flying through the Cahuenga Pass in a Jet Ranger helicopter. It was the last leg of his then-daily route to deliver "bank paper" to a waiting Lear jet at the Hollywood Burbank Airport. (The Lear jet would then fly to Chicago, the final destination.) Alan was talking to the Burbank airport control tower, which was "just around the corner," so to speak—the tower said is was "severe clear" at the airport. There were, however, low-level broken clouds moving in around the pass, but the Hollywood Freeway was brightly illuminated from the traffic moving along, even at one A.M., so there was seemingly sufficient visibility to navigate through the pass. He had a passenger with him: a "bag man" for the bank paper. This man wasn't a pilot, and had a quite obvious fear of flying. When recalling the flight, Alan said he looked over at the man and then back out of the front of the helicopter: not ten yards away from the windscreen of the helicopter was a string of high-tension wires precisely at his altitude, with one of those balls directly in front of the windscreen, lit up by the helicopter's landing light. The large orange balls mounted at various distances on the high-tension wires are there to alert pilots of the wire's presence, but sometimes helicopters approach right between these balls and consequently don't see the balls, only becoming aware of the wires when they are extremely close. Alan's helicopter was a second or two away from impact. He said he jerked the control stick straight back, pulling the helicopter into a vertical climb, and at the same time kicked the pedal to do a maneuver akin to what's called a hammerhead. He heard a loud thump, but the helicopter was still flying, so he maneuvered the ship level and moved sideways over a patch of grass next to Cahuenga Boulevard. After hovering for short time to regroup and stabilize the chopper, he set the machine down on the grass, got out, and looked it over, finding nothing wrong. (The thump turned out to be a loose extra can of oil in the rear cargo area.) He picked up the ship, flew back to the building roof in downtown Los Angeles from which he had departed, and landed. The speechless "bag man" was still in a state of shock. There's a famous paper that many aviation enthusiasts read religiously called *Trade-A-Plane*. One of the perennial advertisements in it reads: "Your Balls Saved My Life"—an ad for those orange balls on the high-tension wires that, thankfully, saved Alan's life that night. Sadly, a friend of Alan's died some time later, hitting those very same wires—in broad daylight. At a 120-plus miles an hour, you don't have to be distracted for long, looking down, for example, to find yourself entangled in the wires. Over the years Alan has lost three fellow pilots to that very same circumstance.

Helicopters also play a role in another frequent source of accidents that I investigate: heli-skiing. Helicopter skiing is as fun and chic as it looks in those glossy magazine ads, but it is also dangerous. Heli-skiing presents several dangers not present at sea level. The first, and perhaps most obvious, is bad weather, which means snow, ice, and visibility problems. The second is altitude. High altitude is a problem for aircraft because it's more difficult to generate lift at higher altitudes. This problem is exacerbated with an attendant issue: typically heli-ski operators fill the helicopter to capacity with passengers—it's expensive and people want the most for their money. So, you have a heavy (sometimes overloaded, in fact) helicopter flying in extremely challenging conditions. Landing on very uneven terrain is another problem that has bitten more than a few heli-ski operators as well. Typically, insurance on a helicopter is more than fifty to sixty thousand dollars a year, and it's obvious why. Those rates also make it quite challenging to realize a profit in operating a heli-ski company.

Those issues are what led my ex-husband (to remain unnamed in this book) and me long ago to purchase and operate a snowcat, in lieu of leasing a helicopter, for back-country skiing. Being an avid skier myself, including having taught skiing and even run a back-country ski operation for years, has made me hyper-aware of the dangers of flying in mountainous terrain, especially in the wintertime. I loved to ski out in the Sierra back country near a town called Bridgeport, California, just north of Mono Lake. Back in the 1970s, Disney had plans to start a ski resort there, but it never materialized. My ex and I had skied the back country via our family snowcat for years and knew the terrain like the backs of our hands. A snowcat is like a big, enclosed truck that has tracks on it, which gives it the appearance of a tank.

It was nearing the end of the ski day one day in the 1990s and we had decided to take one more run. But the weather began moving in much more quickly than forecast, which very often happens in the mountains. When we reached the ridge below Dunderburg Peak (elevation 12,374 feet) where our snowcat driver would typically drop us off, we were engulfed in clouds. The ridge we were on was above tree line and all of a sudden we were stuck in a whiteout. The wind picked up, which made things even worse, as any evidence of our snowcat tracks in the rocky and snowy ridgeline had now vanished. We all gathered around the snowcat and realized that we needed to forgo our planned ski run and get down into the trees where the visibility should improve, and we could make our way safely out of the back country. My ex, who often referred to himself as "the pathfinder," immediately exclaimed that he knew exactly which way it was back down and pointed in that direction. Something didn't sit right with the idea of just heading out in a direction when we couldn't see anything. So I dug my compass out of the pack I wore. I needed to stand back from the snowcat, as the compass did not work correctly when I was in it or near it, due to the heavy metal construction of the vehicle. When I read the compass I realized that not only was my ex mistaken about which way to go, he was off by 180 degrees! The complete opposite direction! I ended up skiing right next to the snowcat driver's side window, where he kept the window down so that I could read the compass and guide us to tree cover. It was very slow going, cold and snowy as could be, but eventually we made it below the tree line. It was like another world: we were protected from the winds and below the clouds now, so we could see again. We made it safely back home. It was a good thing we had ventured up in a snowcat and not a helicopter.

Those conditions are precisely the ones that have often caused many heli-ski helicopters to crash: high winds, low to no visibility, and extremely challenging terrain. Heli-skiing on a bright sunny day is one thing, but in those conditions it would have very likely resulted in disaster. Weather, as we all know, is dynamic. In the mountains it's very dynamic and can go from good to horrible in a matter of minutes.

While many of the cases I've worked on involve accidents, weather and crime make an especially interesting twosome. There is a folk saying that when the Santa Ana winds begin to blow in southern California, the "crazies" come out. Santa Ana winds are very strong, dry, downslope winds, usually easterly or northeasterly, that can sometimes be very hot. This is a deadly combination for fires, both natural and man-made, and it never seems to fail that some arsonist tries to start a fire during these conditions. Over the years, I have worked on numerous fires that occurred during Santa Ana wind conditions. Once the fires begin, it is nearly impossible to put them out due to how quickly the strong winds spread the flames. Firefighters are hindered in putting them out because the Santa Ana conditions make it difficult to operate aircraft—frequently helicopters—for firefighting purposes.

But the hot weather along with the wind also brings out violence in some people.

Hot weather has long been associated with increased violence. The United States, for example, experiences significantly higher crime rates in cities when the temperatures rise appreciably above the city's average. The relationship between heat and assault as been verified, whereas the correlation between heat and homicide is more complicated. Some areas, e.g., Los Angeles, Cleveland, Puerto Rico, and Texas, have been shown to have a positive correlation between hot temperatures and homicide, while others do not (New Jersey is one example). Interestingly, extreme rainfall also leads to increased conflict among people, as is seen in India. Some countries, like developing countries in Africa, are more susceptible to conflict and wars when agriculture is threatened by drought conditions and hot temperatures.

Leaving children or pets locked in cars as the temperature rises: is this a crime? Sometimes it is just forgetting while other times it is intentional and an attempt to cover up a killing by disguising it as an accident. Using the weather as a weapon or to commit a crime is more common than one might immediately think. Jan Null, CCM, is a colleague who has done some wonderful work bringing to light the dangers of hyperthermia in children left in automobiles. He analyzed fatalities of children left in vehicles over a fourteen-year period and discovered that 53 percent were forgotten by the caregiver, 29 percent were playing in the vehicle unattended, and a shocking 17 percent were intentionally left in the vehicle by an adult. In these circumstances, determining whether the deaths were intentional or not involves investigation by authorities. The majority of the deaths were of children under two years old. He also found a dramatic increase in deaths of children left in cars after the advent of airbags, as now children are placed in backseats instead of the front because of the risk to children from the airbag. Vehicles are particularly dangerous in the heat because even cracking the windows has little effect on the temperature inside the car. Within just ten minutes, the interior temperature in a closed car can rise almost 20 degrees Fahrenheit, and after an hour the temperature rises approximately 43 degrees F inside the car as compared to outside. So if it is 77 degrees F outside, after just an hour it will be 120 degrees F inside the car!

FOUR

The War Effect

At anchor in Hampton Road we lay,
On board of the Cumberland, *sloop-of-war;*
And at times from the fortress across the bay
The alarum of drums swept past,
Or a bugle blast
From the camp on the shore

Then far away to the south uprose
A little feather of snow-white smoke,
And we knew that the iron ships of our foes
Was steadily steering its course
To try the force
Of our ribs of oak

Down upon us heavily runs,
Silent and sullen, the floating fort;
Then comes a puff of smoke from her guns,
And leaps the terrible death,
With fiery breath,
From each open port.

—Henry Wadsworth Longfellow
from his poem "The Cumberland"

Around 340 B.C., Aristotle wrote *Meteorologica* (Latin for "meteorology"), a treatise of four volumes that included theories about earth sciences, which encompassed rain formation, water evaporation, and other weather phenomena. Though it contained many errors—such as the following description of lightning: *when there is a great quantity of exhalation and it is rare and is squeezed out in the cloud itself we get a thunderbolt—Meteorologica* was considered by many to be the authority on meteorology for two thousand years.

Weather lore also played a major role in the development of meteorological theories and ideas, and such lore still persists today. *The Farmer's Almanac* still exists and many people love to watch Punxsutawney Phil each year. Some weather lore, like our friend Phil, is just that—lore—but there are other instances of weather lore that is actually based on fundamental weather theory, such as the aforementioned case of "Red sky at night, sailor's delight/Red sky in the morning, sailors take warning." Other examples of sound weather lore are the many about seagulls and coming rain and storms.

Seagull, seagull, sit on the sand,
It's never good weather while you're on the land.

When I was attending Lincoln Junior High School in Santa Monica, California, we had a large grassy field surrounded by our running track. The sand running track encircled more than two acres of grass. Many of the classrooms looked out onto this field. Every time a rainstorm came into southern California, the seagulls would land on our school's grassy field about twenty minutes before it would begin raining—every single time. There were dozens of them. They would all just stand around on the field making squawking noises at each other. Then as it began raining, they would hunker down and just hold still with their feathers all puffed out so that they appeared twice as big as they really were. Then, just as suddenly as they arrived, when the rain showers moved out of the area, they all flew away and went back to wherever they came from, most likely the beach, as it was only a mile away from our school.

There may not be a scientific explanation for the seagulls' behavior. It's an interesting anecdote about the nature of nature and its creatures' behavior. It is theorized that seagulls land in these safe areas when rain is coming to avoid being blown out to sea, and to avoid swaying and possibly breaking trees, wires, or potentially dangerous perching areas.

Historically, weather lore alone is not enough for accurate weather forecasting. While it is has always been important for agriculture and trade, there is another reason weather forecasting is such a social priority: war. It is war that prompted the United States to begin official weather forecasting in 1814, when the U.S. Surgeon General began to collect weather observations from various army posts. From these data came a national network of observers to monitor said data, which then became a true national weather observation network. This national network then became the U.S. Weather Bureau and eventually the National Weather Service (NWS), as we know it today. But it was during the twentieth century that the field of meteorology began to advance rapidly, in particular due to the increasing demand of accurate forecasting for the military. In fact it advanced so rapidly that there were new discoveries almost daily. Radar was one of those huge advances in the twentieth century. At first radar was used to track ships and aircraft, and then it was used to detect precipitation. As the twentieth century quickly became the most martial century in recorded history, new technology was invented to aid in war efforts, which aided in a better understanding of various weather theories, especially in the arena of weather forecasting. The first computer model used to forecast the weather was created in the 1950s. Then, in 1962, the first meteorological satellite images from the TIROS-1 (Television Infrared Observation Satellite) were used in weather forecasting. It was

NASA's first experimental step to show that low-earth orbital satellite's could be used for weather observations.

Weather itself has shaped the outcome of many battles and wars over the years. Onsets of fog didn't allow ships to sail during the Napoleonic Wars, and those like the *Endymion* became lost in fog. Bitterly cold weather killed many troops, as in Napoleon's doomed invasion of Russia in 1812. Plentiful rainfall made soldiers unable to travel during in the Vietnam War. So if weather can be accurately forecast during battles, this can greatly enhance the chance at victory for the side with the most accurate forecast.

D-Day stands out dramatically as the most important weather forecast in history. There are many theories on what the "D" stands for in D-Day (many say it stands for nothing), but what is nearly universally agreed upon is that it began the climactic battle of World War II. Codenamed Operation Overlord, the battle began on June 6, 1944. This day began the Battle of Normandy, which lasted into August 1944 and resulted in the Allied liberation of western Europe from control by Nazi Germany. On June 6, 1944, D-Day, more than 160,000 American, British, and Canadian troops invaded a fifty-mile stretch of the beaches that run along the Normandy coast of France. Before embarking on the battle, General Dwight D. Eisenhower reminded the troops that "the eyes of the world are upon you."

Years of preparation went into the Battle of Normandy. The Operation Overlord chief meteorologist was Captain James Stagg of the British Royal Air Force. He received knighthood in 1954 and became Sir James Stagg, in recognition for his contributions. The Allied commanders knew that their landing on Normandy beach required a full moon and that it needed to peak just after a low tide. These conditions were necessary in order to illuminate obstacles in the water, including underwater defenses installed by the Germans, and to be able to spot their landing locations prior to the break of dawn. The landing boats needed to precede a low tide so that all of the underwater obstacles installed by the Germans could be blown up in advance. But then they needed the tide to begin rising so that they could then get their amphibious crafts onto the beaches. Complicating this further, each of the five beaches along the coast had different low-tide times and thus different landing times. In addition to the moon and tide requirement, the right weather was also required—good weather!

The Allies could not risk losing landing craft to stormy seas; the boat commanders would also need good visibility for landing on the beaches and the air support would also require good visibility. Wet, soggy ground was undesirable, as it would slow down the troops' movements once they made landfall. In short, the Allies required a rather complicated weather menu.

The Axis powers were as cognizant of the weather as the allies. General Field Marshal Erwin Rommel, known as "the Desert Fox," had been tasked by Adolf Hitler to build underwater and beach defense structures against the Allies. This building became what was termed the Atlantic Wall. This wall spanned 2,400 miles along the northern coast of France, both on land and along the shoreline. It contained obstacles, bunkers, and land mines. Because of this wall, Rommel, who was well aware of the timing of the tides along the Normandy beaches, again and again told his troops of the anticipated invasion, "When they come, it will be at high water," thinking that the Allies would try and float over the wall. But this was not to be the case. When the tides were mid-level or higher, the wall was submerged and not visible, though still dangerous—hence the Allied desire for low tide and the visibility that came with it.

General Eisenhower had chosen June 5, 1944, as the first of three possible days in the narrow window of astronomical conditions required. The vertical range from low to high tide always exceeded twenty feet. During low tide at Normandy, large rock outcroppings and portions of the beaches were exposed. Most important, the Atlantic Wall was exposed, allowing the Allied boats to destroy it in advance so that it would not rip up and sink their delicate landing craft.

Talk about pressure to produce an accurate weather forecast! The beaches along the Normandy coast face due north and are only 100 miles south across the English Channel from the British coastline, and 150 miles from the center of London. The weather in this region is fickle and can be tricky to forecast due to the convergence of various air masses from the Scotland, Wales, and England landmasses, and then the open waters of the English Channel. This means that it can turn cloudy and foggy quickly, much to the chagrin of the locals, even to this day. On June 4, 1944, meteorologist Captain James Stagg had the unenviable task of telling the Allied commanders that D-Day was not going to happen on June 5, as General Eisenhower had planned. There was a three-day window from June 5 to June 7 when the astronomical conditions were what they required; if the weather didn't cooperate, they would have to wait weeks for their next opportunity. Stagg's meteorologists, however, disagreed on how long the stormy weather would last. In fact, the American meteorologists thought that June 5 would be clear weather and the stormy weather would not actually come into play until the sixth. Thankfully, Stagg held his ground and told the commanders to delay twenty-four hours. There was considerable disagreement among the American and British forecasters as to when the weather would clear. Knowing the British weather better than his American counterparts did, Stagg had the final say. Utilizing weather stations in Ireland and on board ships, Captain Stagg and his team of British meteorologists predicted that there would be a brief respite in the weather on June 6. He reported this to General Eisenhower, and D-Day was a go for that day.

History has proven Stagg right—for the most part. There was a brief break in the weather on June 6 but the weather was still not ideal. It was cloudy with rough seas, causing some beach landings to miss their intended marks, but by midday the weather became clear and sunny. Aiding the Allies was the fact that the German meteorologists had forecasted the weather turning stormy and remaining stormy for many, many days, and thus did not think the Allies would be staging their attack on the sixth. The Luftwaffe's chief meteorologist reported this forecast to the Nazi commanders and they actually withdrew many troops from their posts along the Normandy beaches. The weather stations that the British meteorologists relied upon turned out to be crucial for providing a more accurate forecast for the invasion, which helped the Allies catch the Germans by surprise. The Germans were correct in their thinking that for the remainder of June, the weather would turn worse, and thus the D-Day invasion might potentially not have occurred at all that month. But the Allies took advantage of that narrow window when the tides, moon, and weather came together perfectly and helped set the stage for western Europe's liberation.

The Korean War was also fought under some extremely difficult meteorological conditions, and accurate forecasting played a major part in strategy. When American troops first arrived in Korea in July of 1950, the weather was hot and sticky. Then, as the months marched on, the weather became cold, rainy, and then snowy. If slogging through mud from the extreme rainfall events wasn't enough, the brutally cold weather in Korea was, in many ways, another enemy the soldiers had to battle. The soldiers lost fingers, toes, feet, whole hands, and even their lives due to the extreme cold weather.

When the Vietnam War began, weather forecasting also played a pivotal role.

"We had a red phone. Yes, it was actually red, by our bed," Katherine Westmoreland told me. She was the wife of General William Westmoreland, and they were friends with my paternal grandparents, Jean and Wilson Williams. We were at a dinner party and I was asking all about the trials and tribulations of living in Vietnam and the very difficult decisions that General Westmoreland had to make during the war. Mrs. Westmoreland and her daughters were just lovely, and we talked about their experiences abroad. Though I talked with General Westmoreland, too, I felt more comfortable asking questions of Mrs. Westmoreland.

"Did you ever answer the *red* phone?" I asked.

"Oh no," she replied, telling me that was only for her husband to answer. "We had another phone that we used for other calls, but when the *red* phone rang it meant there was an extreme emergency that needed his immediate attention," she continued.

"Did the red phone ever ring?" I was so curious and I am sure this came through in my question.

"Oh, yes!" she stated emphatically. "And usually in the middle of the night, of course."

Her husband was a four-star general in the U.S. Army and commander of the U.S. Military Assistance Command in Vietnam. Throughout the course of his command, military strategy and weather forecasting would converge just as they had for Eisenhower a generation before. Weather forecasting and the use of weather stations were critically important. The Vietnamese weather station observations were deemed useless by the

Americans, as many of them were coded in a way that rendered them unintelligible. Other times, the data were falsified, and other facilities just didn't work. To date, the Vietnamese weather monitoring stations are a hodgepodge of observation stations obtained from many countries. In fact, the United Nations, along with Vietnam, is funding an implementation of updated and strengthened weather forecasting and early warning systems to help the Vietnamese bring their weather stations current. In the absence of reliable weather station data during the war, satellite imagery—a new technology for this war—along with reports from fighter and reconnaissance pilots were critical for gaining weather observations over Vietnam.

The first weather satellite, TIROS-1, was launched from Cape Canaveral, Florida, on April 1, 1960. It was now possible to monitor the Earth's weather on a regular basis. TIROS-1 was launched into space and was a polar-orbiting satellite so it was in low earth orbit, or about 530 miles above the earth's surface. However, TIROS-1 stopped working just a few months later, in June of 1960, so TIROS-2 was launched on November 23, 1960, to take its place. Though the images were a bit grainy, one could see things like typhoons, large storm systems, and cloud bands closing in on particular areas of interest. By the mid to late 1960s, NASA, the National Oceanic and Atmospheric Administration (NOAA), the U.S. Weather Bureau, and the Department of Defense (DOD) all had a variety of operational weather satellites. These satellites were getting better and better by the year. Now meteorologists could make out individual clouds and types of clouds, and could interpret wind flow. Then, by the 1970s, the satellite technology from the Department of Defense, their Defense Meteorological Satellite Program (DMSP) was finessed even further, and individual lighting strokes could even be detected on the nighttime images.

Satellite imagery became one of the most important tools for forecasters during the Vietnam War, as the weather in North Vietnam was often extremely cloudy, making visibility and gathering data by any other means nearly impossible. Weather would not have affected just air strikes but also the troops on the ground. It was hot and humid and the heavy rains, along with the incessant presence of insects, caused travel by foot almost to cease. Ground leadership needed to know when the rain would clear so they could move troops forward again, or else they wanted to know how long the rain would also be thwarting their enemy. The Americans even resorted to using weather modification during the war for just that purpose. It was called Operation Popeye. Highly classified at the time, it involved attempting to extend the monsoon season over the supply routes used by the North Vietnamese and the Viet Cong. Aircraft loaded with silver iodide were sent from a U.S. base in Guam on missions to fly over strategic regions of Vietnam and seed the clouds by dropping the silver iodide into them. Silver iodide is still used today in weather modification programs. This cloud seeding caused increased rainfall, such that the supply routes were rendered useless due to saturated soils, landslides, and roads being washed away. It is reported that these missions extended the monsoon season thirty to forty-five days!

During times of war there are large advances in the sciences, and weather is no exception. These advances are also changing the way wars are fought, whether it is cloud seeding or today's focus on remote-sensing, unmanned aircraft surveillance, and monitoring of the enemy's movements with these same unmanned aircraft that also collect weather data. One of the many defense-related projects that I have worked on was the NASA X-43A project. It involved running an atmospheric model for forecasting and then reconstructing the weather for its initial flights. The X-43A was an unmanned hypersonic research aircraft that used scramjet technology to reach speeds of up to Mach 10. Conventional turbojets, like the ones used on a commercial airline, operate by compressing air with fanlike blades that push air into a compressor and then mix this air with fuel to produce thrust. In ramjet engines, a stream of air is compressed by the forward speed of the vehicle and mixed with the fuel. Normal ramjets operate with subsonic internal airflow speeds to create combustion. However, at times the airflow reaches maximum speeds of Mach 2 to Mach 5, as combustion chambers overheat at higher speeds, whereas scramjet technology employs ramjet engines but operates with supersonic internal airflow speeds (faster than Mach 6.7), and will not overheat. Not having to carry its own oxygen for combustion, scramjet engines require half the weight at liftoff. The X-43A aircraft was carried on the B-52 bomber along with a booster rocket. The

B-52 didn't have bombs, but instead carried the X-43A up to high altitudes (approximately 95,000 feet) where the X-43A then started its own engine and flew off under its own power.

On November 6, 2004, the X-43A set a world speed record for a jet-powered aircraft of Mach 9.6 (nearly 7,000 miles per hour!), tripling the top speed of the SR-71 Blackbird aircraft. The X-43A program's greatest challenge was to rely upon and confirm the concept of air-breathing hypersonic flight. The current world record for the fastest unmanned aerial vehicle is the U.S. Defense Advanced Research Projects Agency (DARPA) hypersonic scramjet HTV-2 Falcon, which went 13,201 miles per hour (Mach 17+) on April 22, 2010.

The new technology advances over the past few decades have been heavily focused on developing unmanned aircraft. Personally as a scientist, I still believe in manned missions, although I understand the obvious positives of unmanned aircraft in military engagements. But as far as non-military missions go, manned missions are what get people, especially young people, interested in science, aeronautics, and aviation. It is also why I am so excited about the Perlan Project, which was always focused on a manned glider, where the pilots would harness the energy of the atmosphere to reach very high altitudes. This project encompasses all aspects of a great project: scientific discovery, pushing the limits of manned flight, and as a possible platform for surveillance of our planet, which can be used for science (especially meteorological science), but also for military missions, too. There is a fine line between working on projects that are military projects and others that are defense related, as many scientific advances can be utilized by defense departments with just a few modifications. The atmospheric scientist, and friend of mine, Dr. Joachim "Joach" Kuettner (or Küttner) knew this as well as anyone.

Dr. Kuettner originally obtained his doctorate in law and economics at the age of twenty-one in 1930. Born in Breslau, Germany, Dr. Kuettner was in the midst of the political destruction of his country as the Nazis began to take over in the early 1930s. It was during this time that Kuettner came to the realization that he was more interested in the atmosphere than the law. So he switched gears and obtained a second doctorate, but this time in meteorology. During World War II, Kuettner was commissioned to work at Messerschmitt. There he flight-tested the world's largest glider, the German Messerschmitt Me 323 Gigant. This large assault glider was to be used mainly by the Luftwaffe for transport and could take up to two hundred people onboard, and was intended to be used to invade the United Kingdom. Then came the need for developing a motorized version of the glider. While test flying the motorized version (which had six engines), the Gigant glider broke apart in midair and Joach Kuettner began falling. He had a parachute, but the rip cord was tangled and stuck. He fiddled and fiddled with it and couldn't get it unstuck. Below him he felt a big *whumpf* and realized that it was the Gigant hitting the ground; it immediately burst into flames. Joach was finally able to get his rip cord free and pulled it, deploying his parachute just 660 feet above the ground and landing safely. After this experience Kuettner said all he wanted to do was to go to a mountaintop and be alone to study the atmosphere. This is exactly what he did: for three years, he studied atmospheric phenomena at an observatory at the top of Zugspitze, the highest mountain in Germany. He hated that he was being forced to work on machines that could be used to study the wonders of the sky, but were instead being used to kill people that much more efficiently.

In the early 1950s Joach Kuettner emigrated to the United States, but he made it clear that "I am not going to work on military problems." Joach worked on everything from lee waves to rotors to thermals to hazardous atmospheric winds. He would remain in the United States for the duration of his career, and I feel privileged to have worked with him.

"Elizabeth, you need to learn German," Joach stated emphatically to me one day when we were both attending a convention in Berlin, ironically, as keynote speakers.

"I know. I took it in graduate school but I only feel comfortable reading it," I responded. Joach Kuettner, Einar Enevoldson, and I were all in Berlin as part of a glider convention. Einar and I had given the keynote presentation, and after us Joach took the stage. He blew everyone away with the wonderful story of landing a glider in a farmer's field . . . and this farmer had a beautiful daughter. Kuettner had to spend the night at the farmer's house. The next day, when he went back to town, everyone chuckled about the rumors regarding Joach and the farmer's daughter.

When asked about his love of atmospheric research, Joach responded that two traits kept him actively involved in the field up to his death at 101 years old. These were "curiosity and joy of adventure. If you can preserve these two wonderful afflictions through your life, you will never be able to stop exploring the atmosphere." Dr. Kuettner was the true definition of a Renaissance man. "Curiosity and the joy of adventure" is a pretty good prescription for anyone to take, I would say.

FIVE

The Art of Clouds

> *A cloud is made of billows upon billows upon billows that look like clouds. As you come closer to a cloud you don't get something smooth, but irregularities at a smaller scale.*
>
> —Benoit Mandelbrot

It was with great anticipation that I waited in the audience to finally meet the man who had inspired my senior thesis work at UCLA. I was in my senior year of undergrad and in he walked, wearing glasses, with thinning, frizzy, graying hair, and rather large ears. "Yes, it *is* him!" I thought to myself. It's Dr. Benoit Mandelbrot. This is the man who coined the term *fractal*, which is a complex pattern that is self-similar across different scales, or more simply, a fractal is a pattern that is repeated over and over and repeats in smaller and smaller scales. Nature is full of fractals—mountains, seashells, coastlines, rivers, and, yes, clouds, all contain fractal patterns in their construction.

Fractals are described by their *fractal dimension*. A fractal dimension is a number, and it describes the "roughness" of a shape. Every fractal shape has a fractal dimension. This roughness of fractal shapes is the same whatever the scale. For example, looking at the coastline of California from space, you see that it has many twists and turns; there is the San Francisco Bay Area, then Monterey Bay, then the coast wiggles around more as you head south. Zooming in to another scale and looking at the Monterey Bay/Carmel region of the coastline, it appears the same as the entire California coastline did from space; that is, it jigs and jags with points that jut out with many inlets and little bays. We could keep zooming in until we were at the same scale as an ant walking along the coastline; the ant would see even more jigs and jags in the rocks and pebbles that comprise the coastline. Simple examples of fractals that have repeating patterns are spirals, from broccoli to seashells.

Fractals can be quite beautiful, which is no surprise, as they are some of many instances where art and science are inextricably intertwined, whether it is the earliest cave paintings depicting highly astute observations of the natural world or the Renaissance masters using geometry and perspective to make their paintings more pleasing to the eye. My UCLA undergraduate advisor, Dr. George Siscoe, was the first person to get me thinking about clouds and art. Why do some paintings of clouds look natural and appear so real, while others look, well, terrible? I decided to find out for myself. I spent a year photographing real cumulus clouds around the western United States and then photographing some of the great (in my opinion) cloud paintings. I chose to focus on cumulus clouds—the puffy clouds that often look like pieces of floating cotton, often with flat bases and rounded tops—because, with the help of Dr. Siscoe, we determined that their fractal dimension could be calculated most directly. The nature of cumulus clouds makes them especially good for calculating their fractal dimension. To the eye they can look like cauliflower—which is also a self-similar item whose fractal dimension can be calculated. Once this fractal dimension is calculated, it is fixed. This means that the fractal dimension of real clouds can then be compared to paintings of cumulus clouds. I had the photographs of the paintings and of the clouds blown up to large poster size. From these large posters I was able to make careful measurements and calculations. The closer the painting's clouds' fractal dimension is to real clouds' fractal dimension, the more realistic the clouds appear, as they come ever closer to mimicking the ones in real life.

Some of the great paintings of clouds have been done by painters such as Claude Monet (French, 1840–1926), Jan van Goyen (Dutch, 1596–1656), Odilon Redon (French, 1840–1916), Jacob van Ruisdael (Dutch, 1628–1682), Salomon van Ruisdael (sometimes spelled Ruysdael, uncle to Jacob, Dutch, 1600–1670), John Constable (British, 1776–1837), and Johannes Vermeer (Dutch, 1632–1675), just to name a few. I spent a year from my rooftop, ocean view, non–air conditioned UCLA office measuring and calculating the fractal dimension of cumulus clouds. I then compared the fractal dimension of actual cumulus clouds to the fractal dimension of many of these painted cumulus clouds to determine which matched nature most accurately, and if there was any correlation between the ones that were the most aesthetically pleasing and the ones that had the closest fractal dimensions to real clouds.

During my research, I discovered that the seventeenth-century master painters of the Dutch "Golden Age" were some of the best painters of meteorologically accurate clouds. This Golden Age spanned from 1575 to 1725, with most of the master painters working from 1609 to 1672. The new Dutch Republic was the most prosperous nation in Europe, leading in science, art, and trade, and this era continued until the French invasion of 1672. Today, there is a Cloud Appreciation Society, formed in the United Kingdom, and touting almost 40,000 members in 111 different countries. Some of these members have actually debated whether the Dutch seventeenth-century masters used real clouds as their models. They argued about whether the clouds they painted were modeled on real clouds they viewed at the time, or whether they distorted them or made them up in order to fit their compositions. I would guess that these artists spent countless hours observing and drawing and painting clouds from real life in order to achieve the level of accuracy that they did.

Through my research, I determined that of these painters of clouds, the very best painter of natural cumulus clouds was Jacob van Ruisdael. The fractal dimension of his cumulus clouds best matched those of natural cumulus clouds. Though all of these master painters were wonderful, the purpose of this research was to determine which of the painted clouds matched natural clouds the best and to shed light on why some clouds look more *realistic* than others, and it all boiled down to fractal dimensions.

But what *is* a cloud, anyway? A cloud is visible to the naked eye and contains tiny water droplets and/or ice particles and it floats in the atmosphere above the earth's surface. A cloud can also be formed of another sort of particulate matter in the atmosphere, as long is it is dense enough to be seen by the human eye—for example, a dust cloud. If a cloud touches the ground, it is called fog. The term "smog" used to mean a combination of smoke and fog, but today it more commonly refers to air pollution, usually urban, that may or may not contain natural fog, too.

Clouds come in all shapes and sizes. In fact, there is such a wide variety of clouds that there is a cloud classification chart one can use to determine the name and type of cloud. Clouds appear white because the cloud droplets and/or ice crystals are large enough that they reflect all wavelengths of light (i.e., colors), which, when combined, are seen as the color white by our eyes. But at times when clouds are extra thick or blocked by other clouds, they appear gray instead of white.

Various elements dictate whether a cloud is formed or not. These include temperatures at various altitudes, the amount of water vapor, the wind, and the interaction of air masses. Also required is dust or particulate matter, enabling the droplet or ice particle to form droplets, as most of the time "spontaneous nucleation" of the cloud droplets or ice crystals does not occur as it takes too much energy. Most of the time the droplets form using a particle as the nucleus.

Nineteenth-century British pharmacist Luke Howard was fascinated with the weather. An amateur cloud enthusiast, he was the first person to classify clouds. He proposed three basic cloud types that we still use today: cirrus, cumulus, and stratus. Cirrus is Latin for "curl" or "tuft;" these are the high-level wispy clouds that sometimes look like horse tails. Cumulus is Latin for "heap;" these are vertically growing clouds, and were the focal clouds for my painting-versus-nature fractals study. The third type of clouds are stratus clouds, which is Latin for "layer." These clouds often look like a thick gray or white blanket, very low in the sky, and they often mean rain (or snow).

Clouds are classified similarly to flora or fauna, and have moved beyond Howard's three basic types. Today, there are four families of clouds defined by the altitude within the troposphere: high-level, mid-level, low-level, and vertically growing. These low-, middle-, and high-cloud categories vary depending on whether you are near the polar regions, temperate regions, or tropical regions on the earth. In general the high-level clouds are around 20,000 feet, middle ones are at 6,500 to 20,000 feet, and low-level clouds are from the ground up to 6,500 feet above the ground. There are then ten genuses, fifteen species, and a variety of sub-species and special forms. For example, there are the lenticular or mountain wave, UFO-looking clouds (altocumulus lenticularis); there are hole punch clouds (cirrocumulus stratiformis undulates); and contrails, or clouds of ice crystals formed behind jet aircraft (cirrus fibrates).

Reading the sky is a wonderful skill that can be learned by anyone at any age. My favorite clouds are the unusual cloud formations such as cloud streets, ship tracks, mammatus, lenticulars, and noctilucent clouds, which are depicted in photos in this book, since a picture is worth a thousand words. There are many cloud observer books and field guides for the beginner or expert cloud watcher that show photographs and describe in even greater detail these and other cloud formations that you can look for right outside. In fact, you can learn a lot about the atmosphere by reading the sky. The foehn gap (sometimes written as föhn gap), for example, is a gap in the clouds on the lee side (or just downstream) of a mountain range, usually forming in between cloud layers on the upstream side of the mountains and the lenticular clouds (they can look like space ships) further downstream of the mountains. The foehn gap shows you where a region of sinking air exists; then the air begins to rise again farther downstream of the gap, where the clouds begin to form. This is a sign of mountain wave activity, as it is the gap between the airflow coming down from the mountains, into the valley, and then rising again.

While I was learning all about clouds and the atmosphere at UCLA, we had a TV weather forecaster as a guest speaker one day. His name was Dallas Raines. I have always been curious if that was a stage name or his actual given name. I found out that Mr. Raines had a bachelor of science in broadcast journalism and earth science and had taken graduate-level courses in meteorology. A few days after Dallas Raines's presentation, some of us in the department were discussing TV weather personalities and their various experience in terms of whether they knew anything about meteorology or not. They varied greatly from just news/weather readers to actual meteorologists. But I did not get a clear picture of who was a meteorologist and who was not until years later, when I became involved with the American Meteorological Society (AMS).

What exactly is a meteorologist? Is it the forecaster at your local weather service office or the somber officer in the military uniform giving the weather forecast on the nightly news (yes, this is done in Italy) or the buxom lady reading the forecast on the morning news program? Well, the answer is, it depends. It depends on their level of education and experience. Unlike many professions, being a meteorologist does not require a license that is regulated by state laws. There is, however, a definition of a meteorologist for the United States and there are certification programs.

The American Meteorological Society was founded in 1919 and is a professional organization for the atmospheric and related hydrologic sciences. The American Meteorological Society defines a meteorologist as follows:

> *A meteorologist is an individual with specialized education who uses scientific principles to observe, understand, explain, or forecast phenomena in Earth's atmosphere and/or how the atmosphere affects Earth and life on the planet.*
>
> *This specialized education is most typically in the form of a bachelor's or higher degree in, or with a major or specialization in, meteorology or atmospheric science consistent with the AMS Statement on a Bachelor's Degree in Atmospheric Science.*
>
> *There are cases where an individual has not obtained a degree in meteorology or atmospheric science but has gained sufficient knowledge through coursework and/or professional experience to successfully fill professional positions, such as military weather forecasters or positions typically held by degreed meteorologists. These individuals can also be referred to as meteorologists. This includes*

> *individuals who have obtained and maintain either the AMS Radio or Television Seal of Approval or the AMS Certified Broadcast Meteorologist designation.*
>
> *Individuals who have little formal education in the atmospheric sciences, and who disseminate weather information and forecasts prepared by others, are properly designated weathercasters.*

The AMS describes what is required for a bachelor's degree in atmospheric science but leaves an out for those who do not have a college degree in this subject or a related field. This is where calling oneself a meteorologist gets a bit sticky. Showing that one does, or does not, have "sufficient knowledge through coursework and/or professional experience" is open to a wide variety of interpretations and depends largely on who is judging the credentials of the prospective meteorologist.

The AMS has various certification and seal of approval programs. The Certified Broadcast Meteorologist (CBM) assures that the broadcaster is a meteorologist as defined by the AMS and not just reading a script from a teleprompter. There was an older program called the AMS Seal of Approval, but it was overbroad in its definitions and not as rigorous as the current CBM program. As of December 31, 2008, the older program is no longer offered, although I still see some TV weathercasters using this out-of-date credential, which only required a demonstration tape of the person forecasting in action, with very few meteorology courses required. Many of the seal holders write "Certified Meteorologist" after their name, which is true, albeit still open for interpretation. It is the lowest level of certification with the Certified Broadcast Meteorologist (CBM). The next level of certification has specific educational requirements including a degree in atmospheric science/meteorology or an appropriate equivalent field of study.

The highest level of certification by the AMS is the Certified Consulting Meteorologist (CCM). This program was established as a service for the general public so that anyone hiring a professional meteorologist that holds a CCM knows that this person has "established high standards of technical competence, character (ethics), and experience for certified consultants who provide advice in meteorology to the public." The CCM program was established in 1956 to ensure that certain individuals have been tested and found to meet or exceed these standards and thus companies or individuals could expect a standard of trust and professionalism before hiring someone. I have been a CCM since February of 1997 and was awarded CCM number 572. Yes, at that time I was only the 572nd person to obtain a CCM in the entire United States. When I received my CCM, I took the list of all CCMs awarded and calculated that about 3 percent of CCMs were women. Having served on the CCM Board as well as having been chair of the Board, I know this number is higher now, but still not many women are CCMs. A little more than seven hundred CCMs have been awarded now (including both genders), but fewer than three hundred are currently active, as many have retired or died since the CCM process first began. The very first CCM was awarded in 1957 to Henry T. Harrison. Henry Harrison studied weather conditions on arctic expeditions, was a meteorologist in the U.S. Weather Bureau, and then spent more than thirty years working for United Airlines as that company's chief meteorologist.

The U.S. Weather Bureau that Henry Harrison worked for eventually became what we know today as the U.S. National Weather Service (NWS). The NWS is the only entity that can legally issue weather warnings for the general public. Other governmental entities may issue warnings particular to aviation and boating, but only the NWS can issue warnings for the full gamut of weather, including, but not limited to, heavy snow, high winds, flash flooding, thunderstorms, and tornadoes. In the past, some television "meteorologists" would issue their own weather warnings that would sometimes differ from the NWS warnings and cause great confusion (not to mention potential public safety risks). In South Africa, independent weather forecasters (i.e., forecasters who do not work for the South African Weather Service) can face up to ten years in prison or a fine of up to 120,000 U.S. dollars if they issue incorrect severe weather warnings without official permission.

There have even been lawsuits about overblown weather forecasts in various countries. Most lawsuits regarding inaccurate or inadequate weather forecasts filed against the U.S. federal government or individual states are resolved in favor of the government because of the Federal Tort Claims

Act or the particular state's Tort Claims Act. This law allows forecasters to use their own judgment or discretion when issuing weather forecasts. Many private weather companies have also been sued in the past, usually unsuccessfully due to weather forecasting not being an exact science. Perhaps the presumptions and presuppositions that other sciences are exact and therefore weather science should be as well, are the result of these lawsuits. Weather is not an exact science. But the truth is, other sciences also similarly flawed, as our understanding of the world, whether it is biological, chemical, or meteorological, is always evolving as new studies are conducted and new discoveries are made and then interpreted. Like the paintings I studied all those years ago, there's a lot of art to science, and it isn't limited to clouds.

SIX

It's Super Cool to Supercool

A man who carries a cat by the tail learns something he can learn in no other way.

—Mark Twain

I didn't see it at first—it was dark as coal outside. But as our snowcat drove us closer, I could begin to see its outlines against the night sky. Then, there it was: Storm Peak Laboratory, my home for the next month in the winter of 1988, at 10,500 feet above sea level atop Mount Werner in Steamboat Springs, Colorado.

I had thought we'd never make it. We had spent the past two days trekking across Nevada and Utah. After leaving Reno, we broke down in Winnemucca, Nevada, better known for hookers and bars than for meteorological treasures. After spending a night in Salt Lake City, we headed over the mountain passes—our truck pulling a huge trailer full of scientific instruments, computers, a snowmobile, and tools—before finally making it to the front range of the Colorado Rockies.

When the other scientists and I arrived at Storm Peak Lab, known less formally as SPL, we had just enough supplies in our backpacks for one night. We were too tired, and it was too late, to bring anything else up to the lab, which was located at the top of the Steamboat Springs Ski Resort. Other equipment and supplies would have to wait until the morning.

It was about 22 degrees Fahrenheit as we climbed out of the snowcat and thanked the ski resort employees for the lift up the hill. The lab was dark, and as I entered I saw that it was really just two mobile home trailers hooked together by a room, which I would later learn was known as the "cold room." The lab had no running water and no flushing toilet, but it did have heat, electricity, and a chemical toilet, similar to the kind one would find in a porta-potty. The cold room and the roof of the lab were where I would be spending most of my time and, not surprisingly, I was going to love every minute of it.

The next morning was beautiful, bright, and sunny. I dressed in my ski gear—the only way to descend the 3,668-foot drop to the bottom of the mountain was on our snowmobile, pHred (pronounced "Fred"). The name was apt science jargon, given that pH, the measure of the acidity or alkalinity of the clouds and snow, was one of the many things that we were there to measure and study. Plus, pHred was red.

The only ways up at the time, in the late 1980s, were via snowmobile, snowcat, or ski lift. But because the lift didn't go all the way to the top of Mount Werner, taking it meant we had to hike the final half mile on our skis. (Today, there is a lift to the top of the hill that lets off near SPL, which is in a brand-new, beautiful, fully functioning building that stares out over the Yampa River valley.)

Though other scientists would come and go to perform their own experiments or just to visit, our core team for the next month was Dr. Randolph "Randy" Borys, Dr. Edward "Ward" Hindman, David "Dave" Mitchell, and me. Randy, Dave, and I were from the Desert Research Institute in Reno, Nevada, and Ward was from the City College of the City University of New York (CUNY). It was the winter of 1988 and I was a graduate student working on my master's degree in atmospheric physics at the University of Nevada at Reno, with which DRI was affiliated. Randy (along with Dr. Joseph Warburton) was my advisor, and Dave Mitchell was a staff researcher. Dave is now Dr. David Mitchell, and Ward is now professor emeritus at CUNY. The DRI took control of the lab from Colorado State University in 1989 and Dr. Borys, the lab's new director, has big plans for it, including the construction of a new building at this mountaintop location.

The next morning, we hitched a trailer to pHred and brought the remainder of our gear up to the lab. It would take a couple of days to fully unload it all. We began setting up our instruments, equipment, and computers alongside the many computers and scientific instruments that were already in place and ready to go. Then it was time to hurry up and wait: we were there to collect cloud water, snow crystals, and precipitation particles and to sample the air during winter mountain storms. We didn't have to wait long.

"Did you see the forecast?" shouted Dave. "The storm is headed right for us and should be here in about thirty-six hours!" During our wait we had made sure that everything was ready: we had placed the atmospheric instruments and other instruments on the roof of the lab. One of the instruments we used was a one-of-a-kind device we called a "snow separator." It resembled a casket. Dave had designed the snow separator and the guys at the shop at DRI had built it for him. Along with our separator, there were also the mounting devices for the "cloud sieves" that Ward and Randy had developed; a classical scattering spectrometer probe, or a device that shoots a laser beam between two metal "eyes" that allows us to measure the sizes of clouds droplets; and many other instruments with fancy names: a cascade impactor, a cloud condensation nucleus spectrometer, a Royco particle counter, and even a "cloud gun" droplet spectrometer, all of which are designed to measure and count various particle sizes and cloud droplets, depending on the instrument.

Down below us, the ski resort was quite busy, as it was a Saturday. Occasionally, brave souls would hike up to a place called "the toots" to get some out-of-bounds skiing, and on their way up they would have to trek by our lab. If they hadn't seen SPL before, they would look with wonder at the alien-like lab, with its instruments sticking out all over the place. I was shoveling a path in the snow to the main entrance of the cold room when three male skiers approached. One of them asked me if we had a bathroom and if he could please use it. I couldn't help but detect the urgency in his voice. I told him that, yes, we had a chemical-type toilet that he could use. I led him through the cold room and into the trailer with the toilet. In front of the toilet hung a flimsy curtain, to give the illusion of privacy.

When I got back outside, a fitter-looking man with bright blond hair asked me, "What *is* this place? I have always wondered what goes on here." I introduced myself. The man said his name was Steve. The dark-haired guy did not seem that interested when I began to explain that we studied things such as pollution, clouds, snow, and climate change. Steve took off his skis and I took him on a tour of the lab, leaving his buddy behind. We started with the cold room; then I took him to the side of the lab that had the computers and the panoramic view—eliminating a tour of the toilet for the sake of some decorum.

"You see, up here we are on the western edge of the Colorado Rockies in what we call 'free-stream' airflow," I said, as we looked out of the large picture window down to the valley below. "That means that as storms approach from any direction—the northwest, west, or southwest—we can actually sample the clouds and precipitation to determine where the air came from. We analyze the chemistry of the clouds, the snow crystals, and the aerosols. Aerosols are the tiny airborne particles on which cloud droplets or ice crystals form. They can be natural particles, such as dust, or they can be pollution spewed into the atmosphere by cars, industrial plants, consumer products—you name it. So when you analyze the chemistry of the air, the clouds and the snow falling from them, you can tell what they formed onto."

I decided to stop there, figuring that was about as much information as an anxious skier could patiently absorb. I was surprised, however, that he kept asking questions. He blurted another one out: "But just analyzing the chemistry doesn't tell you where it came from, does it?" Coming as bit of a surprise to me, he was actually *thinking* about what I had been explaining to him. Many more people are now taking an interest in what's happening to our environment because we all have pretty much become witness to the constant controversy over climate change in the media and see extreme climatological events on a regular basis. Steve, however, seemed more informed than most.

His questions were quite probing, so I continued, "Certain chemical constituents we can automatically attribute to specific industries or sources. We analyze for all sorts of things such as sulfate, nitrate, chlorine, and ammonia, just to name a *very* few. Take sulfate: in the atmosphere, it results from the burning of fossil fuel, the combustion of biomass. That includes basically anything that burns: oil, gas, forests, agricultural fields, or wood. Sulfate and nitrate are two of the main players in acid deposition, which used to be called acid rain. At times we even perform what are called 'tracer studies' to narrow down a sample's possible emission sources. To do that, we look at a particular storm period and basically backtrack, going back days or even weeks to determine where the air came from. Then we look for particular chemicals that we can attribute to a specific place—perhaps based on the type of dust that's found there—or to a specific industrial source. Scientists have traced pollutants collected here at the lab to sources in the Pacific Northwest, the Los Angeles region, and even Canada. One time there was a pollutant that scientists couldn't attribute to any American or Canadian industries, and it turned out to be from industrial plants in Beijing, China! Quite remarkably, the pollutants traveled sixty-three hundred miles to get here!"

Steve still wanted more. He gathered his thoughts and finally asked, "But *how* do you *actually collect clouds*?" I answered him as we made our way from the inside of the trailer to the rooftop, where he could see more of our instruments. "Well, you see, water is really an amazing and highly misunderstood and underrated substance. For one thing, it can 'supercool.' This means that it can remain liquid at temperatures well below freezing *without* freezing. In fact, cloud droplets can, at times, remain in a liquid state all the way down to minus forty degrees Fahrenheit." (That happens to be where Fahrenheit and Celsius meet, thus it is also –40 degree C; remember, water's freezing temperature is 32 degrees F.) I paused, thinking about all of the other amazing properties of water, including the fact that water can exist in all three phases (gas, liquid, and solid) at its triple point (a specific point of temperature and pressure), but I didn't want to go off on a tangent with curious Steve still standing there in his ski gear. I continued on, "But in this supercooled state the droplets are very unstable. This means that if they hit anything—a fence, the wing of a plane, a ski lift, or the goggles you are wearing—they will freeze on impact, forming ice. The cloud droplets are very, very small. In fact if you pluck a hair off the top of your head, the thickness or diameter of your hair is about ten times thicker than an average cloud droplet. So in order to collect them, we stick these tennis racket-looking devices called cloud sieves up on the roof of the lab. When the droplets freeze and build up on the sieve's 'strings,' *voilà*, we have just collected cloud!"

It was getting late, and I had to finish shoveling and prepping for the imminent storm, so I led Steve down from the roof and through the cold room to where he had left his skis and his buddy. By this time his other friend was done using our toilet and was already in his skis ready to go. I said good-bye to Steve and he thanked me profusely for the unanticipated science lesson he had just received. I let him know that he could stop by any time whenever he wanted. I never did see him again—some great skiing and a hot toddy at the lodge probably won out over having to use a chemical toilet and another science lesson—and I can't say that I blame him! Nevertheless, I am always excited to teach people a bit about weather and climate, given the heightened awareness of what is happening in our environment today.

It was seven P.M. on this January day and the storm we had been waiting for was just beginning to hit. We were all doing our own predetermined tasks, like preparing our instruments and our snow, cloud droplet, and cloud particle collection areas by making sure all of the work surfaces were sterilized, as the wind began to blow harder and harder outside. It was about 12 degrees F outside, and with the wind chill it felt more like –10 degrees F. We were nearing conditions in which one can get frostbite on any exposed skin after just ten minutes. I was wearing my full gear: *two* pairs of long underwear, thick ski pants, a turtleneck shirt, *two* sweaters, a ski jacket, a hat, ski gloves, goggles, and my prized Sorel boots. I was ready for battle!

I began what would soon be my daily routine. I climbed the metal ladder that went from the inside of the cold room to the roof of the lab. That climb became a real chore: as the night went on and ice and snow accumulated from the boots tracking it in, the rungs got very slippery. I headed up the ladder with the first of six cloud sieves and installed it into the mounting device on the roof. It was an amazing sight up there. Huge, bright lights shone from every direction, beautifully illuminating the mist of the clouds that fully enshrouded us.

I shared the ladder and the cold room during those times with Dave, who tended to the snow separator and collected samples of snow crystals that fell out of the cloud. The heaviest snow crystals, those that had collided with and become covered by cloud droplets, fell in the first collectors; the lightest, those barely touched, or not touched at all, by cloud droplets, fell in the bins farther down inside the "casket." Not only do the snow crystals collide with the cloud droplets, they also collide and sometimes stick to each other, forming huge snowflakes. Unfortunately, I didn't have time to admire the beauty of it all as I had five more cloud sieves to put up on the roof before the storm began in earnest. I had to get going.

The cloud cover was still light, giving me time to get my routine down pat before I lost more visibility. Up on the roof, there were six sieves of varying size and weight. Each sieve collects droplets of a certain size range, which we call size segregation. The smallest-stringed sieves collect the smallest, and most prevalent, cloud droplets, so those fill up first. As the strings go up in size, to the large tubes used in the largest sieve, they collect larger droplets. We are then able to determine if the chemistry of the cloud droplets changes with droplet size.

It turns out that the chemistry varies greatly. Why does this matter? Interestingly, clouds are very efficient cleaners of the air. They accumulate chemicals such as sulfate, nitrate, chlorine, potassium, sodium, and ammonium, and eventually deposit those chemicals on surfaces (trees, the ground, etc.), on ice crystals, and into snowflakes or raindrops that fall to the ground. Cloud droplets contain more chemicals, by mass, than snow or aerosols, and the smaller, more prevalent cloud droplets are also the most concentrated chemically! Not all of those chemicals are pollutants; some droplets also form on natural particles from the earth or the ocean. But most of the chemicals are due to human activities. There is a good side and a bad side to this cycle. The good: many pollutants are "cleaned" out of the air we breathe by these droplets. The bad: they can get washed into the soil, aquifers, lakes, and oceans, where they can affect humans, animals, and plants from below.

One of the benefits of the science done at SPL is that it helps unravel mysteries of climate change and potentially enable us to create solutions, or at the very least, better understand what is happening. Global emissions from humans (called anthropogenic emissions) from cars, trucks, and other vehicles, from the burning of biomass and biofuel, and from various manufacturing industries have a profound impact on the Earth's climate system. Pollution does not obey political boundaries; it invades via the oceans, water, and the ground, and is free to travel around the globe via the atmosphere. A cloud does not care where the pollution came from, but we, as humans living on this beautiful planet we call earth—breathing her air, using her resources, drinking her water—should care deeply.

The scope of the droplet research we were doing at SPL even extends beyond understanding the long-term causes and future implications of climate change. For example, the fact that water can supercool is sometimes a matter of immediate life and death importance: supercooled cloud droplets have brought down many an aircraft. How can tiny droplets cause an aircraft to crash? They can build up on the aircraft frame, in particular the wings and tail section; they can get into carburetors (that is, if the aircraft has a carburetor) and cause it to choke and quit working; or they can build up on the pitot (pronounced *pee-toh*)-static system—the system of pressure-sensitive instruments that measure vitals such as airspeed and altitude—and cause it to display incorrect information. Ice crystals, not supercooled cloud droplets, can also sometimes clog up the pitot-static system. All of these are called "icing" events.

Two cases that illustrate the dangers of airframe icing and icing of the pitot-static system come to mind: the crash of Colgan Air Flight 3407 and the crash of Air France Flight 447. The Colgan Air flight, a commuter flight for Continental Airlines, departed Newark, New Jersey, headed for Buffalo, New York, on February 12, 2009. On its approach to the Buffalo-Niagara International Airport, the Bombardier Dash-8 Q400 aircraft began icing up. Supercooled liquid water was freezing on impact as it struck the wings, tail, and other critical surfaces on the aircraft. The aircraft had a system of rubberized "boots," located on the leading edges of the wings, tail, and propellers, that are supposed to inflate and "pop" off any accumulated ice.

There are essentially two types of systems in use to deal with ice on aircraft (there are others, but those are not relevant to cite here). The first is the leading edge rubber boots, which represent very old technology and often produce somewhat unpredictable and unreliable behavior when inflated. If the boots inflate asymmetrically, or one doesn't inflate at all, it will have a pronounced effect on the aircraft controllability. This system is referred to as the de-icing system—meaning the pilot of the aircraft waits until a certain amount of ice accumulates on the surfaces and at that point inflates the boots with air in the hope that the ice will delaminate from the affected surfaces. This system has yet another problem when it comes to supercooled raindrops, and that's called runback icing. This occurs when the ice accumulates in areas *behind* the boots, where it's not possible to rid the aircraft of it. It can quickly add substantial weight and drag that will quickly impact the aircraft performance drastically—maybe even cause a crash, as it did on the flight to Buffalo.

The second type of ice management for aircraft is called anti-icing. This is used on aircraft with critical wing design that cannot tolerate any ice accumulation. This system uses what's called "bleed air heat" from the engines and routes it to the leading edges (and engine inlets) on the aircraft. It is supposed to be turned on (usually around 40 degrees F) when there is visible moisture, before entering conditions that may produce icing. This system is typically used on jets and high-performance aircraft and is considerably more effective and reliable than the de-icing boots, which are more common on smaller, commuter-type aircraft. From my experience as a forensic meteorologist reconstructing many icing conditions that have brought down aircraft, the best system is to avoid icing situations altogether, as neither in-air de-icing or anti-icing system is foolproof.

Regarding the Colgan Air crash, the National Transportation Safety Board concluded that the captain responded inappropriately to the icing and the "stick-shaker" alert that warned him of an imminent stall. This is a tactful way of saying that the pilot screwed up. He in fact did exactly the opposite of what he was supposed to do: he and his copilot pulled back on the controls, pulling the nose up further and exacerbating the stall situation, instead of pushing forward to drive the nose of the aircraft down and get out of the stall. The aircraft stalled, plunged to the ground, and crashed into a house, killing all forty-nine on board and one person in the house.

On June 1, 2009, the Air France Airbus A330, Flight 447 was on its way from Rio de Janeiro to Paris when it plummeted 35,000 feet into the middle of the Atlantic Ocean, killing all 228 people on board. What could have caused such a tragic event?

I can only imagine what the passengers were thinking and doing during those final moments when, inevitably, they must have realized they were going to die. Today when I am flying and the weather is bad, I think about what I might do were I in that situation. Would I have time to turn on my cell phone and call my loved ones? Probably not. Would I be able to scribble a quick note to my family telling them how much I love them? Maybe. I have worked on cases where these sorts of notes have been found. If I were working on my computer and using in-flight wi-fi, I would be able to use Apple's FaceTime (or Skype or one of the other video conference programs). I frequently use FaceTime when I am traveling in order to stay connected with my husband, Alan, and our ten-year-old son, Evan. It seems to ease the pangs of separation, especially for Evan. Sometimes the passengers sitting next to me on a flight make funny faces at the screen and play around with my son. He loves it. However, I've decided that videoconferencing would be terrible, since it would mean my loved ones would actually watch my demise. I'd opt for the written note. Hopefully, no one has to consider these options, as each is terrible, although I have a feeling I am not alone in thinking these morbid thoughts every time a plane crash makes the news. But I digress.

Air France Flight 447 crashed due to a combination of weather and piloting mistakes. The Airbus 330 headed directly into the strongest portion of a line of severe thunderstorms. Why?

After thousands of man-hours by teams of experts, the story of what happened to Air France Flight 447 is now known. It took more than two years, until May 2011, to find and recover the aircraft's black box from the floor of the Atlantic Ocean. Early on, many rushed to judgment and speculated that the crash might not have been an accident but a bombing. As soon as I looked at the weather charts and satellite imagery, however, I noticed that the aircraft had crashed shortly after entering a region of severe thunderstorms. Of course, it immediately occurred to me that the accident might have been weather, not terrorism, related. Other aircraft had been avoiding that particular region of clouds and thunderstorms, but Flight 447 flew directly into it. I had seen enough of these cases, and was familiar enough with the consequences of thunderstorms (i.e., icing, turbulence, lightning, and hail), to know it likely wasn't a coincidence that the aircraft crashed after entering this region of severe weather. I was retained to work on the case not too long after the accident.

I found that the aircraft had indicated incorrect airspeeds because the pitot-static system had clogged up with ice crystals and perhaps some supercooled liquid water. Ice crystals were probably the main culprit, because the outside air temperature, about –40 degrees F (–40 degrees C) at 35,000 feet and –62 degrees F (–52 degrees C) at the aircraft's cruising altitude of 39,000 feet would have been too cold for supercooled liquid water to exist. On the Air France aircraft, there were two pitot-static tubes, each about four inches long, which stuck out on the sides of the aircraft near its nose. Designed to let air enter into the pitot-static system, those tubes iced up and became clogged, causing the pilots to receive incorrect airspeed and altitude information. Though the tubes are heated, the heating elements couldn't keep up with the formation of ice crystals as the plane flew into the severe thunderstorms, precipitation, and turbulence.

The copilot was flying the aircraft, and about two hours into the flight, he began to have trouble with his instruments. The aircraft began to roll and the stall warning triggered briefly. He called for the captain, who arrived one minute and forty-seven seconds later. They both struggled to figure out why they had lost control of the aircraft. They never considered that the pitot-static system might have iced up and been giving incorrect information. Indeed, they never formally identified the stall situation. The autopilot, which is driven by the air data computers (which receive information from the pitot-static system—the pitot tubes) on board the aircraft, was not receiving correct information and disconnected. Whenever the pitot-static system senses a failure of some sort, it will automatically disconnect the autopilot. The pilots, now hand flying the aircraft, proceeded to make a critical series of mistakes in assessing what was happening, and literally flew the plane, which now was in a full stall, right into the ocean. More than four and a half minutes passed from the moment the plane began having trouble to the time of impact. That is a *long* time for the people on board to endure. It's heartrending to see a tragedy like that occur when it could so easily have been prevented. Even after venturing into thunderstorm conditions, if the pilots had broken the stall—which could have been done by correctly analyzing what was happening to the aircraft and adjusting airspeed and angle of attack accordingly—they could have simply flown the airplane out of the stalled condition and at a sufficiently low altitude to get through the storm.

Over the years, I've learned a lot about storms and icing and my experience at the Storm Peak Lab stands out as one of my favorite first hands-on forays into the physics of these amazing phenomena. They're powerful and nothing to be glib about when you're at 39,000 feet. I can remember bringing down my first cloud sieve during that first big winter storm at SPL. It was, of course, the smallest stringed sieve, which collected the smallest cloud droplets and therefore iced up before the others. In fact, during parts of the storm the largest sieve took up to an hour to accumulate enough droplets to make for a decent sample. I climbed down the ladder one-handed, holding the sieve in the other hand and trying not to fall head over heels into the cold room. (Later I discovered a technique for climbing down single-handed: I would put each foot into the corners of the ladder struts to secure my footing on each step. Try this, should you ever need to do it yourself.) I set the sieve onto a sterilized plastic sheet on my workbench. Unfortunately, I then had to take off my warm, protective ski gloves and put on sterilized, freezing cold, non-insulated rubber gloves. My hands were sweaty from climbing up and down the ladder, and it would certainly have been a tragicomic act to watch me struggle to put those gloves on. After I finally got them on, I grabbed a sterilized plastic scraper and began scraping the ice off every string on the sieve. I had to make sure I got every corner, an exercise that quickly became tedious. I needed to get a move on; the storm was picking up and the other sieves needed attending to. I labeled the plastic sheet with the time, sieve number, and some other codes, wrapped the plastic tightly with a tie, and put it in a cold-storage box. There was no time to pause and wonder where that particular sample came from. I climbed back up the ladder, replaced the smallest sieve, and took the second smallest sieve back down into the cold room.

And so it began. As the storm rolled in, the sieves began to ice up faster and faster. No sooner did I complete the largest sieve than I had to immediately go back to the roof and gather up the next one. As the sieves iced up more and more rapidly, I developed a better ice scraping technique. But I still couldn't keep up with the most intense part of the storm. I laughed to myself and recalled the episode of *I Love Lucy* in which she and Ethel were working in a chocolate factory. Their job was to dip the candy in the chocolate, but they had trouble keeping up with the chocolates as they sped by on the conveyor belt. Their boss, nonetheless, decided that the belt wasn't speeding by fast enough and yelled out, "Speed it up!" Unable to keep up, Lucy and Ethel began eating the chocolates as fast as they could and then stuffing the rest in their clothes, including their hats. That was how I felt that day up on Mount Werner.

Finally the storm passed and we were done sampling. It was by then almost eight A.M. the next day. We had gone for thirteen hours straight. I was tired and starving, but euphoric with the success of my first "battle." One thing I knew for sure: Research is tedious, hard, gratifying, frustrating, disappointing, and extremely fulfilling, all at once. I fell fast asleep. I'd just carried a cat by the tail and I was too tired to even eat.

SEVEN

A SAGE Experience

You can know the name of a bird in all the languages of the world, but when you're finished, you'll know absolutely nothing whatever about the bird . . . So let's look at the bird and see what it's doing—that's what counts. I learned very early in life the difference between knowing the name of something and knowing something.

—Richard P. Feynman

It was 96 degrees F outside, I was wearing a light pink dress and low heels as I walked into the concrete bunker of a building . . . in this place where the graduate students were 90 percent male and 90 percent were of foreign nationality. The concrete structure was one of twenty-four of these SAGE (Semi-Automatic Ground Environment) buildings constructed across the United States. This decommissioned building now housed the

Desert Research Institute, where I worked on my graduate Master's thesis research and then worked on achieving my PhD immediately thereafter. The University of Nevada at Reno, where I would do my coursework, and the Desert Research Institute (DRI) are closely affiliated with each other. The DRI SAGE building was located north of Reno in a town called Stead, Nevada. You could see it from miles around. It was that foreboding, decommissioned SAGE building where I would be spending almost the next decade of my life.

The twenty-four SAGE buildings were constructed across the United States as a result of the Cold War, and they were designed for one reason: to house a huge IBM Whirlwind computer and key military personnel. The walls of the three-story building were three feet thick, and devoid something I usually enjoy looking out of: windows! The building had only one air source to the outside and it was quite small, only about two times the diameter of a vacuum hose. Using the "canary in the coal mine" technology quite literally, the military placed a cage full of birds in each room, and if these birds became sick or began to die, the air from the outside world be immediately cut off and the buildings would begin running their own internal air system. This procedure was implemented so that the U.S. military and essential parts of the government could still function in the case of an attack. All of the SAGE buildings were connected via a "crosstelling" communication network and interfaces with the Army Air Defense command posts and missile radar networks. These SAGE buildings had many other functions, including searching for incoming missiles and also for guiding Bomarc missiles and tracking radar in the vicinity. The SAGE system was in operation until 1983 when it was decommissioned as the threat of the Cold War no longer seemed imminent.

In fact, portions of Stanley Kubrick's film *Dr. Strangelove* were filmed at the SAGE building in Nevada where I was working. Inside the place I was often reminded of a line in the film, George C. Scott (playing the gum-chewing, paranoid General Buck Turgidson) was wrestling with Peter Bull (who played the Russian ambassador, Alexi de Sadesky). In the midst of their scuffle, President Merkin Muffley (Peter Sellers) screams at them, "You can't fight in here, this is the war room!" The name SAGE itself could easily have been used in the movie—it sounds like something Terry

Southern (one of the screenwriters of the film) would have come up with. Semi-Automatic Ground Environment definitely has a cinematic ring to it.

It was in that concrete bunker that I would make many new weather and climate discoveries, including developing a better understanding of the water budget of winter mountain storms and unraveling the relationship between cloud droplet size and the chemistry of these droplets, all while enjoying numerous experiences with my fellow colleagues who represented a lively international community. Many of my colleagues came from all over the world, including Eritrea, Turkey, Italy, China, Taiwan, Sweden, Bangladesh, Greece, Croatia, Serbia, and the United Kingdom. The town of Stead, where our SAGE building was located, is quite famous for the Reno Air Races and we all enjoyed watching them from the rooftop of the building whenever they occurred. The SAGE building was positioned for great aerial visibility, given its original military purpose.

Even without the looming threat of the Soviets, the atmosphere is always in a sort of a war. Anyone, for example, who has had a very turbulent ride in an airplane penetrating one of those "beautiful, puffy white clouds" that look so peaceful has had firsthand experience with the war that is always going on within air masses. Warm air masses, cold air masses, water, wind, precipitation, air pressure . . . it all comprises what we call weather, and weather is often churning visibly and invisibly in a battle of its own. Weather fronts are a perfect example of this, whether it be a warm front, where warm air is intruding on relatively colder air, or perhaps a cold front, where the opposite happens.

As is the case with all scientific investigation, research plays a critical role in our understanding of the elements that create weather. I've been involved in several different research projects, some that occur inside a warm facility and some outside and knee-deep in icy elements, like the Sierra Nevada Mountains in the state of Nevada. While there, I often found myself with a rope wrapped around my waist alongside my colleague, Renyi Zhang, as Renyi and I began pulling a sled. A man named Belay Demoz was pushing from behind, all with the goal of moving a snow separator that was lying on top of the sled. Our snow separator was a big, heavy, and unwieldy instrument made out of plywood coated with polyurethane. It was more than eight feet long and almost two feet wide with a large four-foot-tall "chimney" mounted on top. The instrument was designed to separate rimed from unrimed snow crystals and ice particles. "Riming" is the process by which snow crystals and ice particles fall through a cloud, and in the process, tiny cloud liquid water droplets hit these particles, freezing on impact. This greatly affects the chemistry of the snow and ice. The snow separator is also able to separate snow by ice particle type and amount of riming, be it a stellar crystal, a needle crystal, or a rimed graupel particle. Just as with clouds, there is an ice particle identification type chart found in most cloud physics books. The multitude of ice crystal types are formed by variables in temperature and vapor supplies (humidity).

Back in Nevada, more than two feet of freshly fallen snow made hauling the monstrous piece of equipment up to the Tahoe Meadows Laboratory quite a feat. At 8,550 feet elevation, the lab was located off Mount Rose Highway, which connects Reno to Lake Tahoe. It was the winter of 1987–88 and I had just met Belay Demoz a few months earlier. Renyi was from China and Belay was from Eritrea (a country in Africa that lies along the Red Sea just north of Ethiopia; Eritrea actually land-locks Ethiopia from the Red Sea, and has provided a constant source of conflict between the two countries over the years).

We were three enthusiastic new graduate students in the atmospheric physics department. On that day, the clunky snow separator was the common denominator that brought us together for research pertinent to our respective graduate theses and our own knowledge bases.

"Let me take the rope Elizabeth," the ever-thoughtful Belay requested during one of our rests, as he walked from the back of the sled to the front. Belay had an accent such that he pronounced my name "Eliz-a-bet." It sounded great the way he said it. "Renyi, can you take the back of the sled?" Belay asked.

"Oh, no, you two can't do this alone," I protested.

"Once we get to the top of that hill you can pull it again," Belay said, pointing to a small hill just a little ahead. Even the flats were difficult going in the deep snow. As Renyi and Belay neared the hill, there was a dip in the terrain. So they built up speed to make it up the hill. They began running, so the sled with the monstrous snow separator on top of it wouldn't get away from them. In doing so, one of Belay's "moon boots" came off in the deep snow. Eritrea doesn't have snow and this was the first time Belay had seen it *in vivo*. He was wearing newly purchased, après-ski style moon boots, shoes that are not normally used for pushing a wooden behemoth in deep snow, but more for sipping hot toddies after a day on the slopes. Although one moon boot came off in the deep snow, his momentum kept him going: he continued running in the snow with the one boot and one sock-covered foot. Luckily for all of us, it was a bright, sunny day with no wind.

"Ahy, look at my foot!" Belay exclaimed, as Renyi, Belay, and I broke down laughing. I ran back to get Belay's moon boot and handed it to him. Wiping off only a small potion of the snow that was now stuck to his sock, Belay put his foot back in his boot, snow and all. He again refused to let me take part in the pushing or pulling of the sled . . . and we finally made it to the lab. It was an amusing start to the years that we would spend at SAGE researching cloud physics, mountain meteorology, and climate. Belay was, and still is, a very special person and to this day we remain great friends and colleagues.

Belay and I would spend the next five years together, taking classes and sharing an office in the SAGE building, working on our respective research projects. Dr. Belay Demoz is now a bigwig with NASA and the University of Maryland, Baltimore County (UMBC), as he is director of their Joint Center for Earth Systems Technology. He is married to a lovely woman named Tsigereda and they have three beautiful children together. Our time as graduate students at the DRI SAGE building taught us much about the inner workings of the atmosphere and helped us develop the careers we have today.

The weather on Earth is driven by many factors, but the most important is the sun. The sun's energy is not distributed uniformly over the globe, and this creates temperature differences and thus the flow of air (and the water in ocean) from one place to another. This transfer of heat around the globe occurs both vertically and horizontally. In addition, the earth revolves around the sun (unless some of you are still followers of Ptolemy) on its tilted axis, and because of this, almost all regions, minus the equatorial regions, have seasons. Now it gets much more complicated than this, but this is the general overview of why we have weather on Earth. Describing the basics of any field is an extremely difficult thing to do, and doing so in meteorology is no exception.

A pet peeve of mine is the lack of understanding of dew point temperature and humidity by some television meteorologists. Explaining the difference between the two can be tricky so I find it easiest to explain it by starting with how one *feels* the difference between these two types of measurements. First, the dew point temperature is the temperature the air must be cooled to (at constant pressure and water vapor content) for saturation to occur, that is, for the air to reach 100 percent relative humidity. Well, that doesn't really explain what it means to a layperson. The higher the dew point temperature, the "stickier" one feels, as the higher the dew point, the more moisture there is in the air. Okay, so what? To take it a step further, the dew point tells you how much moisture is in the air, whereas the relative humidity tells you how close the air is to saturation, i.e., 100 percent relative humidity. Here is the scenario: "The weather forecast for this morning is for temperatures to be in the high forties and the humidity is around 95 percent," states a TV weather person . . . but in reality, that air will feel *dry* and cool. It can also be hot and moist-feeling but the relative humidity can be low. How can this be? That is where the dew point temperature comes in. The dew point temperature tells you how much water vapor is really in the air and thus if it will feel dry, comfortable, or humid. In a nutshell, if the dew point temperature is above 65 degrees F, it will feel uncomfortable outside, especially as the dew points increase. The worst combination is high humidity along with dew points higher than 65 degrees F. That is why dew point temperatures are so important to meteorologists in describing the weather.

For snowmaking operations at ski resorts, they not only need to consider the air temperature and the dew point temperature but the *wet bulb* temperature. This also gives an idea of how well the human body cools itself through sweating. Wet bulb temperature, however takes things to an entirely new level of complexity—unwarranted unless you plan on making your own snow.

How do meteorologists get accurate information about the actual state of the weather and the state of the atmosphere and thus come up with things like dew point, humidity, and wet bulb temperatures? This is achieved through an ever-evolving combination of measurements and observations. From the seventeenth to the nineteenth century, instruments were developed to measure the temperature, pressure, and moisture in the air, including thermometers, barometers, and hygrometers. Then in the nineteenth to early twentieth centuries, regional and global networks of meteorological surface observations were developed, as weather stations were placed outside to measure atmospheric parameters. But observations made strictly at the surface of the Earth do not show the entire picture when one is trying to forecast the movement of storms, air masses, and jet streams in the atmosphere. One needs a clear picture of the structure of the lower atmosphere, too, not just information about the surface conditions, for accurate weather forecasting, which was all that had been available for the past three centuries.

A giant leap in weather forecasting came in the 1920s when the radiosonde was invented. A radiosonde is a balloon filled with helium or hydrogen that carries a small, lightweight box tied to it that hangs about six feet below the balloon. Most of us would call this a weather balloon. The box is equipped with meteorological instruments and a radio transmitter. The instruments measure parameters like temperature, pressure, humidity, and winds. Today even more parameters are measured in certain locations, such as ozone at the surface and up high in the atmosphere. The balloons are released and they ascend through the atmosphere and get pushed by the wind along the way. As they ascend, they transmit the data from the meteorological instruments to a ground station. These balloons sometimes burst early, but most of the time they make it up to 100,000 feet or higher before popping. Then a small parachute is deployed and they slowly fall to the ground, where the radiosondes are then retrieved and their data analyzed.

These radiosondes provide invaluable information for meteorologists, who then enter the data into their weather forecasting models. Today there are hundreds of these balloons launched twice a day, always at 0000 Universal Time Coordinated (UTC) (or 7:00 A.M. Eastern Standard Time) and 1200 UTC (or 7:00 P.M. Eastern Standard Time) at hundreds of locations around the world. The two launch times are the same so that all balloons are recording at the same time around the world, giving a picture of the structure of the atmosphere twice a day. The U.S. National Weather Service releases more than 75,000 balloons every year, but only recovers 25 percent of them, which they refurbish and use again. The father of a dear friend of mine, Danäe Anderson, actually played an integral part in the advancement of radiosonde technology. This man, Dr. Kinsey Anderson, was a physics professor at the University of California at Berkeley. He told me all about his experiences making measurements with radiosondes of cosmic radiation impacting the Earth's atmosphere. During some of his work he tested some new balloon material that could potentially fly higher in the atmosphere carrying more weight (i.e., more instruments). Dr. Anderson tested this new balloon by strapping himself into a boatswain chair tied to the balloon. (A boatswain chair is basically a board secured by ropes to make a seat.) Kinsey floated a hundred feet above a concrete floor in this contraption taking measurements. He described it as one of the most thrilling yet scary things he had ever done. Kinsey also told me about how the U.S. military used these balloons to spy on the Russians by launching them in Eastern Europe where they would drift downwind over Russia. Equipped with cameras, these new balloons, which were made out of polyethylene material, were used for reconnaissance. He said that the cameras were then parachuted to Earth over friendly lands or water, where they were recovered.

Today there are numerous additional methods and instruments for obtaining measurements of the atmosphere, including satellites, wind and temperature radar profilers, Doppler weather radar, LIDAR (light detection and ranging) and microwave radiometers, to name a few, but the simple balloon method is still used today, almost a century later, and stands out as the first foray of forecasting based on data gathered from the atmosphere, instead of just from the ground.

Another big step in regard to improving the accuracy of weather models and weather forecasting is the use of aircraft-measured data. There is a relatively new system called ACARS, which stands for Aircraft Communications Addressing and Reporting System. This system sends data and information from aircraft to ground-based or satellite-based stations. Many commercial aircraft are outfitted with this system and provide aircraft meteorological data reports. Some of the U.S. airlines outfitted with measuring devices include American Airlines, Delta Airlines, Federal Express, United Parcel Service, and Southwest Airlines. There are also a number of commuter and charter aircraft outfitted with these systems. These systems collect and transmit meteorological data such as air temperature, wind speed, pressure, and turbulence to various ground stations, where forecasters can interpret the data accordingly via their weather models.

After the attacks on the United States on September 11, 2011, the airspace over the U.S. and Canada was closed for two days to all civil aviation operations. This provided a unique opportunity for meteorologists to study the impact of *not* having the aircraft-obtained meteorological data as input into our weather models. It was determined that the accuracy of the U.S. weather forecast models dropped dramatically simply because of having significantly less accurate data to put into the model calculations.

I feel very fortunate that while working at the DRI SAGE Building all those years ago, I met Harold "Hal" Klieforth. Hal, like me, was a graduate of UCLA. There he obtained his master's of science in meteorology. He then became part of the Sierra Wave Project, which took place in the Owens Valley of California in the lee of the Sierra Nevada mountains, and was part of it from 1951 to 1954. This project was funded by the U.S. Air Force. Its purpose was to investigate mountain wave activity, their associated weather, and their impact on possible hazards to aviation. The Sierra Wave Project used gliders to investigate the wave. Later in the project they added engine-powered planes. Hal worked on this project with Dr. Joach Kuettner, the same professor who had urged me to learn German—a small world indeed!

Hal Klieforth joined the Desert Research Institute in 1965 and he worked there until his death. Hal was an impeccable record keeper of meteorological data and had an amazing memory. If I asked him about a weather event from the 1940s at a certain location in the western U.S., he could tell me about it off the top of his head. While we were at SAGE together, he would take me to the second floor of the building where he kept his treasures. This entire floor contained reams and reams of paper documents in piles placed on shelf after shelf. These piles of documents were all over the place, but Hal knew exactly what was where and could go directly to whatever pile he needed to retrieve whatever he was looking for. These documents were weather records from the past half century to that very day. He had everything from weather charts and precipitation gauge data to photographs and journals of weather from his many driving and hiking adventures. When DRI took over the SAGE building, they cut into the three-foot-thick concrete walls and put in a few windows here and there. I mean a few—I think there were only three or four offices with windows in the entire building. Hal's "document floor" was one of them. Unfortunately, during a severe wind storm back in the late 1990s, a window on that floor broke, creating a "storm" of papers and completely upending all of Hal's careful record-keeping. There were so many papers all over the place, it would take years for someone to even have a hope of organizing them. I have no idea what became of these documents when DRI moved from the SAGE building to a new facility in Reno, Nevada. I still cherish Hal's many letters and the notes he gave me regarding various weather events and research projects over the years, especially this one.

14 September 2010

Dear Elizabeth,

Thank you for sending me a copy of the Kuettner Symposium held in Atlanta earlier this year. I attended his 80th or 85th birthday celebration in Boulder to which I contributed several photographs of him in the 1950's including some taken while traveling with him to Norway, Belgium, and Poland in 1958.

*Joachim and I talked about the design of a high-altitude sailplane for stratospheric flights in the late 1950's, but then he left AFCRL for Florida to join the Apollo Project.**

Looking forward to learning about the Perlan Project.

Sincerely,
Hal

** or, rather, its predecessor, the Mercury Project.*

I miss Hal Klieforth and all his enthusiasm about weather. He never stopped learning and wondering about the earth and its environment. There was something new to observe and explore every day. I cherish my memories of the "sage" people I met in the concrete bunker of a building called SAGE—pun very much intended. We never fought in the war room, either!

EIGHT

Water Wars

I cannot command winds and weather.

—Horatio Nelson

Water is perhaps the most precious and, at the same time, misunderstood substance on our planet. Most of the time it is taken for granted, even though it is essential for sustaining life as we know it. The list of indispensable things that require water is too long to completely enumerate here. Some obvious ones are: agriculture, drinking, hydroelectric power, and the production of goods. Not to mention that we would certainly die without it. The average human will die of thirst in just about a week without water, depending on how much water is stored in his or her tissues. Droughts and water shortages have been, and will always be, in an endless cycle around the globe due to the variability of weather and climate, although with climate change, drought patterns are becoming less predictable, and, in some parts of the world, more severe.

Water is used in many ways that are not always obvious. To produce just one ton of steel takes 62,600 gallons of water. To produce a U.S. newspaper for one day uses 300 million gallons. The earth's surface is 80 percent water but much of this is not directly usable by humans: less than 1 percent of the earth's water is suitable for drinking.

The island of Santa Catalina lies just a little over twenty miles off the coast of southern California. Most people just call it "Catalina." It is a popular destination for day sailing trips, hiking, and kayaking. This small island also has many hotels for tourists, as it can be reached by airplane, helicopter, boat, or ferry. The island is twenty-two miles long and ranges from a little over a mile to eight miles wide; it has a year-round population of 4,000. Yet the island visitor count is 850,000 people per year. Severe drought in the western United States has caused the groundwater to dwindle; that, along with the ever-increasing number of tourists to the island, has put it in dire straits. Water on the island is supplied by Southern California Edison (which also supplies the gas and electricity) and Edison has imposed massive restrictions on water use. It is so bad that the tourists are urged to take just *one-minute showers*! The restaurants serve meals on plastic plates and use plastic cups and cutlery just to avoid washing dishes. All house construction on the island is at a standstill because they do not have water to mix concrete. Eighty percent of the water provided to the island by Southern California Edison comes from a desalination plant and the rest from groundwater wells. So what is the problem? Why can't they supply enough water to the island? Simply, there is only a certain amount of water to go around, even with the desalination plant. They are planning to build a second plant, and between the two they can, hopefully and finally, get out from under the perpetual water woes that have plagued the island for years.

Water supplies are stored in reservoirs, but even these can only hold a certain amount of water. In California, where a state-wide drought has made national news for most of 2015, more than 80 percent of its water supply comes from the snowpack in the Sierra Nevada mountains. If there has been a good winter with lots of snow, then as the temperatures heat up come spring, the snowpack begins to melt, running down the mountains in streams and rivers and into waiting reservoirs. Yet the irony is, in record snowpack years there is more water than the reservoirs can hold and in lean years there is not enough water supplied to fill these basins.

Today, at the time of writing this book, there are drastic water supply shortages in the western United States. The state of California is taking steps toward increasing its water supply, trying to model itself on Israel. Much of Israel is desert and for that reason it has pioneered water generation as well as water conservation technologies. At least 25 percent of Israel's water supply comes from desalination. Desalination, also called desalinization, is a process that removes salts from seawater or brackish water, producing fresh water. At the moment, desalination is still an exceedingly energy-intensive process, but a process that has many obvious advantages; in particular, it does not depend on rainfall or the snowpack for water supply. Historically, Israel has been compelled to deal with their very real water shortage issues time and again, and so their solution has been to build the largest desalination plant in the world, located in Kadima, Israel. Israel has also experimented with and used weather modification over the years to enhance rainfall through cloud seeding.

Here in the United States we are now also facing some very real water supply issues, especially in the West. California is especially prone to these sharp drops in water supply, and given the huge amount of agriculture that takes place in the state, the state is moving in the right direction by building and proposing more desalination plants. Presently the only existing ones are in Monterey, Santa Catalina Island, and San Nicolas Island. There are several more proposed in the San Diego area alone as of 2015, and a very large one nearly completed in Carlsbad.

Another approach is weather modification. Weather modification is used for enhancing precipitation from clouds, suppressing hail, and dissipating fog. There is also inadvertent weather modification from pollution spewed into the atmosphere. As long ago as 1932, there was an Institute of Artificial Rain in the USSR that studied modifying the weather. Then there was a major breakthrough in the field of weather modification when a group of scientists was formed at the General Electric Research Laboratories in New York. The purpose of this group was to find ways to combat aircraft icing and radio static while flying through storms. The group was led by Dr. Irving Langmuir, an American chemist and physicist who went on to win the Nobel prize. In 1946, Vincent Schaefer, a member of this group, was experimenting with the creation of supercooled clouds in his lab: supercooled clouds made of cloud water that exists below freezing as liquid water! He used a home freezer, which he lined with black velvet. He then aimed an ordinary light beam inside it and breathed into the freezer. As he breathed into the freezer, the light beam would show little clouds of moisture that formed due to his breath. Then he would try sprinkling all sorts of things into the freezer such as sand, talcum powder, sulfur, and household cleaners, in the attempt to create clouds made of ice crystals. He always kept the freezer at –4 degrees F, but one day when he arrived at the lab he noticed that someone had turned it off. Anxious to get his experiments going again, he needed to cool the freezer down quickly, so he dropped a pellet of dry ice (solid carbon dioxide at –108 degrees F) into the freezer, and immediately he noticed a trail of tiny ice crystals along the path of the dry ice. What had he just witnessed? The dry ice pellet had turned the cloud droplets from his breath (liquid) to ice crystals, which then fell out of the cloud generated by the dry ice. Later, Schaefer tested this out on a real cloud in the Adirondacks, where he dropped dry ice along a path in a four-mile-long cloud, which caused snow to start falling out of the cloud. Now it was not just a laboratory experiment but a reality, and cloud seeding was born.

Cloud seeding, which was used for military purposes in Vietnam, can be done on warm clouds like the ones found in that part of the world, called hygroscopic seeding, and on cold clouds (supercooled liquid water clouds). I spent much of my early career working in the field of cloud seeding and weather modification on cold clouds. I had the good fortune to work with Dr. Joseph Warburton and Dr. Alexis Long, two giants in the field of weather modification, researching winter mountain storms and determining the best portion of the storm to seed clouds for the most benefit, that is, to get precipitation (water) out of them over the desired areas. These studies took place in Australia and the United States. I also determined the water budget, which in this case was an analysis of the fluxes of liquid water and ice mass in the upwind and downwind portions of these mountain storms. Water budgets are determined by analyzing such parameters as cloud tops, bottoms, cloud liquid water content, winds, and precipitation types.

Many of these cloud seeding projects utilized silver iodide as the ice nucleus. The year after Schaefer's freezer epiphany, Bernard Vonnegut (older brother to author Kurt Vonnegut) discovered that silver iodide worked even better than dry ice for cloud seeding. He deduced that the crystal lattice structure of silver iodide was so similar to natural ice that it might mimic ice very effectively once sprayed into a cloud. Silver iodide has a hexagonal structure, as does ice. Both materials' atoms are arranged identically to the positioning of the oxygen atoms, and there is very similar spacing among the atoms. After testing silver iodide on freezer-created supercooled liquid clouds in the laboratory, Vonnegut discovered that he was correct: silver iodide was a most effective nucleus for cloud seeding.

Then he tested silver iodide out on *real* clouds. He found that vaporizing the silver iodide and spewing it into the atmosphere from airborne and then ground-based generators was the most efficient method. One gram of silver iodide can yield as many as ten trillion ice crystals. Soon, ground-based generators were created to seed clouds from the ground, which became the preferred method of seeding the clouds, as using this technique cost around $2.50 per hour as compared to operating an aircraft, which in the late 1940s cost approximately $25 per hour. Amazingly, by 1950 over 10 percent of the land in the United States was under contract to cloud seeding firms. Cloud seeding projects are still numerous through the U.S., especially in drought-prone states such as Texas, California, and Nevada.

I was able to see one of the two locations in South Africa that have a long history of cloud seeding: Nelspruit and Bethlehem. Nelspruit is in eastern Transvaal, South Africa. It is the capital of the Mpumalanga province and lies near Kruger National Park. Bethlehem is in the eastern Free State province. The cloud seeding operations used a unique method, utilizing flares attached to wings of aircraft. These flares were ignited just below the cloud base and they contained small salt particles. These tiny salt particles were then spewed out into the convective clouds as the aircraft flew just below cloud base, augmenting the clouds and enhancing the rainfall. However, using aircraft for cloud seeding adds much expense to these projects, as noted above, so many of them rely upon ground-based operations.

As we took off from the Nelspruit Airport, it felt like my heart stopped beating for a minute. I was not expecting the runway to end where there was such a steep drop-off, just at the same time we were climbing out. It was thrilling. The Nelspruit Airport lies at 9,518 feet elevation and has since been replaced by the Kruger Mpumalanga International Airport for passenger flights. We were flying a private chartered plane that held nineteen passengers plus two pilots. Our small group took up half of the plane. Luckily it had a toilet—nice to have on longer flights. I was with my mother, Catherine, and a group of friends, and we were on a three-week trip traveling all over South Africa. The two pilots in our hired de Havilland DHC-6 Twin Otter aircraft flew from Nelspruit over Swaziland and down to the southeast corner of South Africa to a province called KwaZulu-Natal. It was 1990. When we arrived, I remember that I was reading the Durban newspaper (Durban is the largest city in the province), and it was being reported that the number of deaths had just outpaced the number of births in the province. It hit me how HIV/AIDS is still a big factor in that country.

Prior to arriving in South Africa, many of our group had previously read Alan Paton's 1948 novel *Cry, the Beloved Country*—it is a classic and was a wonderful book to have read prior to seeing South Africa—it made the experience even more robust. The book begins in a remote village in the Natal province, which in 1994 merged with the Zulu bantustan of KwaZulu to form KwaZulu-Natal. Our flight from Nelspruit to the KwaZulu-Natal province took us to the Hluhluwe Airport, which was a nice change from the harrowing departure from the Nelspruit Airport. We were staying at the Hluhluwe Game Lodge for a few days where all of us, including the pilots, enjoyed the place.

Unbeknownst to us, during our stay, our Twin Otter aircraft pilots had spoken with the managers of the lodge and convinced them to plow down the tall grass and weeds in one of their fields to make a runway that would provide a dramatic departure experience for their passengers. So on our scheduled departure day we loaded ourselves and our gear into the lodge vans and began what we thought was a journey back to the airport. After only five minutes of driving, we were suddenly pulling into a big grassy field where our plane sat, waiting for us. Out in the field already were most of the staff of the lodge. They were all so excited to send us off, as this was the very first time an aircraft had departed directly from their lodge. It was a stunningly beautiful day, clear skies and no wind. We took photographs of this moment with our group, and as we were all standing in front of the plane posing, the entire staff from the lodge began singing their South African national anthem, "Nkosi Sikelel' iAfrika." It is still one of the most beautiful songs I have ever heard. Every time I hear it, I still think back on that magical moment.

FLOOD KILLS 155 IN SOUTH DAKOTA; 5,000 HOMELESS, read the headline of the Sunday *New York Times* on June 11, 1972. What made this flood even more significant was that just a few days before, cloud seeding experiments had been performed in the area. The families and representatives of the families of those who died took legal action by suing the federal government. This is just one example of many lawsuits pertaining to weather modification. The suit was eventually dismissed in 1982 for legal reasons pertaining to who can sue the government for work done by a governmental contractor.

"You're stealing our water!" declared the residents of eastern Nevada. This accusation was aimed at the cloud seeding being performed in the Sierra Nevada mountains, which supply a majority of the water to southern California. Without this cloud seeding, the plaintiffs claimed, the water would have fallen naturally in Nevada, and not in the Sierra, where it would be siphoned off to California. The lawsuit was thrown out of court. The task of the judge is to decide if there is enough evidence to continue with the trial, and in this case it was deemed there was not. In fact, storms that affect eastern Nevada towns, such as Ely, do not generally originate from the Sierra; it is an entirely different storm track that impacts eastern Nevada. Areas about 50 to 150 miles downwind of cloud seeding operations tend to see increased, not decreased, precipitation, as was discovered by one of my PhD advisors, Dr. Alexis B. Long. Various types of weather modification have been used for more than one hundred fifty years, including cannons, large fires, and ground generators that spew chemicals into the air, to aircraft that drop chemicals into the clouds. But is this stealing water from those who live downwind? The short answer is no. The reason is that weather modification hinges on the fact that there is an abundance of water vapor in the atmosphere compared to the cloud liquid water and ice particles. Cloud seeding only brings out the excess water, and does not interfere with what would have fallen naturally. Dr. Edward "Ward" E. Hindman, emeritus professor at the City College of the City University of New York, a colleague and friend, researched winter orographic clouds (clouds that form as air is lifted due to terrain) in Colorado over a barrier (mountain range). His research shows that cloud seeding activities should actually *increase* the amount of inflow moisture that precipitates by 1.3 percent, and so consequently, it reduces the amount of outflow moisture by 1.3 percent. Dr. Hindman concludes that cloud seeding activities should not "rob" moisture from the downwind side of the barrier.

The seeding of clouds utilized the principle called the Wegener-Bergeron-Findeisen Process, named after three of the early cloud seeding researchers. It is usually just referred to as the Bergeron Process. This crux of this process is that ice crystals grow at the expense of liquid cloud droplets when the air is supercooled (i.e., brought well below freezing). So what? This means that when Vincent Schaefer dropped that dry ice particle into the freezer, the cloud of liquid droplets immediately turned to ice crystals in the path of the dry ice. But as the liquid is converted to ice and falls out of the cloud, there is still an abundance of water vapor left over. This means that as the air mass moves on to a new location the "snow" originally inside the cloud was not stolen from the cloud, never to be replenished. It is still there, and waiting to fall.

There are some rather vocal conspiracy theorists who talk about the dangers of "chemtrails" and weather modification. Let's begin this story with a description of *contrails*, not chemtrails. Contrails are the condensation trails that are visible following the path of jet aircraft and appear as white cloud lines emanating from an aircraft flying in the sky. Sometimes they are visible and other times they are not, depending on the atmospheric conditions. These trails are composed of tiny ice particles that form around the aerosols that are emitted in the aircraft exhaust, forming a cloud, or contrail. Aircraft engines emit carbon dioxide, water vapor, hydrocarbons, nitrogen oxides, sulfur oxides, and soot. If atmospheric conditions are right, the hot, humid air from the jet exhaust mixes with the cold atmosphere, forming contrails. Under certain conditions these contrails can spread out and remain to form a thick layer of clouds if there is enough air traffic. Also depending on the atmospheric conditions, contrails may last for hours. With air traffic increasing throughout the world, the incidence of contrails is also increasing. In regions where there is heavy air traffic, the cloud cover has increased by as much as 20 percent.

Now, the "chemtrails theory" is as follows. The government is spewing out chemicals and/or biological agents for sinister purposes without telling us. There are many websites dedicated to instructing the general public on how to document these alleged chemtrails and how to investigate them. I have read and watched videos of some convincing sounding, intelligent people and arguments. Don't misunderstand me, I am sure the government does lots of things we don't know about. But here is the bottom line for me regarding chemtrails. It is very simple to prove or disprove whether our government is spewing chemicals on us without our knowledge. All this effort having people go out and look at contrails to determine if they are chemtrails is wasted. What is needed to prove or disprove the theory is to investigate the chemtrails by taking samples of the contrail itself. Though expensive and requiring effort, this can easily be accomplished. Outfit numerous private aircraft with air core samplers and simple video cameras to document the work. Have these launch at various times and locations all around the states, especially near air force bases, where most of this is said to stem from. Then analyze the samples for barium, strontium, aluminum oxide, and other potential geoengineering chemicals. But this research must be done by scientists using the scientific method, as chemicals such as barium occur in the atmosphere from other sources, such as the burning of coal. So there must be air samples taken well away from the contrail and then other samples from within the contrail. Then publish the work for all to see. The chemtrail-phobes could easily be validated or silenced with a bit of relatively simple scientific research—but maybe they prefer to enjoy idle speculation around the fireplace as a form of paranoia-recreation.

While certain groups continue to fear chemtrails, the truth is that water issues present a far more pressing threat. My research into this area has brought me to Melbourne, Australia, which has a population of more than 4 million, and this city illustrates the water problems many places face. Water usage is growing around the world; in Melbourne it is rising by 2 percent each year, while the size of the water catchment basin storage remains the same. Obviously something must be done. Weather modification, which includes cloud seeding, impacts only a small portion of the problem. There is another water problem we are facing and that is the contamination of our oceans. They are being contaminated by pollution and trash, especially plastics. The majority of our trash is now composed of plastic. It is *not* a myth started by a conspiracy loon that we have large gyres the size of the state of Texas composed of plastics floating around in our ocean. (For those who do think it's a myth, you might want to watch the documentary *Plastic Paradise*). There are five main gyres in the world's oceans and other smaller gyres scattered throughout the world. The main five lie in the north Pacific, the north Atlantic, the south Pacific, the south Atlantic, and the Indian Ocean. Water flows into these gyres and whirls around and around until it reaches the center, where it then flows downward about 1,000 feet and then flows back out again, escaping through the middle of the gyre. These gyres can be thousands of miles wide. But unfortunately, plastic is much lighter than water and is thus more buoyant and is unable to flow downward; it winds up floating just at and usually below the surface of the water, remaining in the gyre forever. Contributing to these ocean trash heaps is the fact that plastic production and use is increasing, not decreasing. The problem is thus growing at an alarming rate.

Part of the problem is that these gyres of plastic lie just below the surface of the ocean, so that they are difficult to see with the naked eye or a satellite. These trash gyres have huge ramifications from water pollution to causing the deaths of sea creatures. These garbage patches in the oceans are affecting our sea life, our birds, and thus us. We are at the top of the food chain and are eating these animals that have ingested massive amounts of the plastic. Plastic does not break down and biodegrade like other materials. It is designed to last, and that it does all too well. It can last years and years. For example, a soda pop plastic bottle takes approximately 450 years to decompose!

Water also plays an interesting role in forensics. Of the many shipwrecks that have occurred in the Great Lakes, the wreck of the *Edmund Fitzgerald* is one of the most mysterious. The S.S. *Edmund Fitzgerald* was first launched in 1958 and was named after the president and chairman of the board of Northwestern Mutual Life Insurance Company, the company that contracted the building of the ship. She was 729 feet long and weighed 13,632 gross tons, and until 1971 was the largest ship on the Great Lakes. She spent her days going between Silver Bay, Minnesota, Detroit (Zug Island), Michigan, and Toledo, Ohio, taking her from Lake Superior to Lake Huron to Lake Erie and then back again. She would be loaded up with taconite (a low-grade iron ore) and take it to the steel mills in the Detroit/Toledo areas.

The fateful voyage of the *Edmund Fitzgerald* began on November 9, 1975, from the Burlington Northern Railroad Dock No. 1 in Superior, Wisconsin. The ship was headed to Zug Island. She was loaded up with 26,116 tons of taconite pellets (marble-sized balls) to be delivered to the Detroit area. Under Captain Ernest M. McSorley, a seasoned captain with more than forty-four years of sailing on the Great Lakes, she departed around 4:30 P.M. local time. About two hours into the journey was joined by the ship the *Arthur M. Anderson* (under Captain Bernie Cooper), which had departed Two Harbors, Minnesota, and was headed on a similar route. The two ships were in radio contact, and the distance between the two varied from about ten to twenty miles.

Lake Superior is the second largest lake in the world (behind the Caspian Sea) and is the world's largest freshwater lake by area. The average depth of Lake Superior is 489 feet, with a maximum depth of 1,333 feet. The large area and depth of the lake surface has a tremendous impact on the weather and state of the lake's water. Most of the shipwrecks on the lake happen when nor'easters occur during the fall. The term "nor'easter" refers to the northeast wind that is usually very strong or even gale force; it can also refer to cyclonic storms that occur along the East Coast of the United States, bringing northeasterly winds over the coastal areas, and can bring gale force winds, heavy snow, rain, and rough seas. On Lake Superior alone, there have been more than 350 wrecks, killing over 1,000 people.

The storm that took the *Edmund Fitzgerald* began on November 8, 1975, in the central plains and headed north, directly toward the Great Lakes. It appeared to be a typical November storm for the region, but by the evening of the ninth it had strengthened greatly. There were multiple National Weather Service warnings and modified forecasts from November ninth to tenth. At seven P.M. local time on the evening of the ninth, about two and a half hours into the journey of the *Edmund Fitzgerald*, the National Weather Service issued one of many gale warnings (wind speeds from 34 to 40 knots or 39 to 46 miles per hour) and forecast that the winds would be northeasterly (wind direction is always the direction it is blowing *from*), then shifting to northerly and northwesterly the following day. Then around two A.M., the NWS upgraded the gale warning to a storm warning (winds 48 to 50 knots or 55 to 63 miles per hour), forecasting winds to be northeast and then becoming northwesterly on the tenth, with waves of eight to fifteen feet. So the captain of the *Edmund Fitzgerald* and the captain of the *Arthur M. Anderson* took a northerly route across Lake Superior. There they believed they would be protected from large waves generated by the northeasterly winds, as the ships would be tucked up against the Canadian coastline where large waves were unable to form due to a small fetch (the distance wind can travel over open water).

At three A.M. the winds in the region of the ships was reported northeasterly at 42 knots (48 miles per hour) but by seven A.M. on the tenth the storm started to move across Lake Superior. By the afternoon the winds had shifted and the storm was almost directly over both ships. The winds had "backed" to the northwesterly at 43 knots (close to 50 miles per hour). The *Arthur M. Anderson* reported steady winds at 43 knots and twelve-foot waves. At this time, the *Edmund Fitzgerald* reported to the *Anderson* that it had started to list. With northwesterly winds, i.e., winds blowing from the northwest to the southeast, there was now a long fetch of water the wind blew over, causing large waves to form where the ships were sailing.

The size of the waves is dependent on the wind speed, the duration of the winds, the size of the fetch (the larger the fetch, the larger the waves), and the air-sea temperature difference (the larger the difference, the faster and larger the waves can grow). The NWS underestimated the size and movement of the storm and kept issuing updates to its warnings, increasing the winds and changing its directions. Meteorologists call what was happening that day "chasing the weather." When forecasts are incorrect and not bearing out, newer forecasts are quickly issued and are indicative of the fact that the forecasters are watching the actual weather and adjusting the forecasts as time goes on. With a storm changing as rapidly as the one taking place on Lake Superior that day, there was a lot of chasing being done.

By late afternoon on the tenth, Captain McSorley of the *Fitzgerald* contacted another ship, the *Avafor*, out on the waters, stating that they had a bad list, had lost both radars, and were taking heavy seas over the deck in one of the worst storms Captain McSorley had ever seen. The *Anderson* reported that it had started to snow, the huge waves were causing sea spray, and at 5:20 P.M., reported the winds were now coming in from 305 degrees (northwesterly) at a steady 58 knots (67 miles per hour) and gusting to 70 knots (81 miles per hour) with waves of eighteen to twenty-five feet! Captain Cooper of the *Anderson*'s description of what happened next is chilling. At close to seven P.M., their entire ship was engulfed in a monster of a wave that plowed down on the ship and traveled all the way along from bow to stern, pushing the bow (nose) of the ship down into the sea. Miraculously the ship popped back up out of the sea only to have another, possibly even larger, wave come crashing down onto the ship. Then Cooper stated that he saw these two waves traveling down toward the *Fitzgerald*, and he believed that these are what sent it to its doom.

The *Edmund Fitzgerald* sank around 7:15 P.M. Eastern Standard Time on November 10 about seventeen miles north-northwest of Whitefish Point, Michigan, near the southeast corner of Lake Superior, killing all twenty-nine men on board. No distress signal was sent from the *Fitzgerald* and analysis of the wreckage shows that the ship probably sank so quickly that the lifeboats were not even accessed.

The wind in the wires made a tattletale sound
When the wave broke over the railing
And every man knew as the captain did too
'Twas the witch of November come stealin'
The dawn came late and the breakfast had to wait
When the gales of November came slashin'
When the afternoon came it was freezing rain
In the face of a hurricane west wind

When suppertime came, the old cook came on deck
Sayin' "Fellas, it's too rough to feed ya"
At seven P.M. a main hatchway caved in
He said, "Fellas, it's been good to know ya"
The captain wired in he had water comin' in
And the good ship and crew was in peril
And later that night when his lights went out of sight
Came the wreck of the Edmund Fitzgerald

—portion of the lyrics of *The Wreck of the Edmund Fitzgerald* by Gordon Lightfoot

Shipwrecks are not a thing of the past—as we are well aware with the recent sinking of the *El Faro* in October of 2015 near the Bahamas as it was trying to navigate through Hurricane Joaquin.

The incredible confluence of events on Lake Superior that led to the sinking of the *Edmund Fitzgerald* is just a microcosm of the incredible and complex interactions that take place between the ocean and the atmosphere. The ocean influences the atmosphere in a number of ways through water evaporation and thus providing energy (heat) to the atmosphere, and consequently the surface of the ocean cools. The atmosphere, in turn, influences the ocean in many ways, such as through winds that blow across the surface, creating waves and currents; and when the winds are strong enough, they cause water to spray, producing tiny droplets of ocean water to fly into the air. Some of these evaporate, becoming microscopic grains of salt that clouds and fog may form later. The ocean currents are well known to scientists. The ocean is full of eddies, or circular currents of water, including the five main gyres that exist in the ocean.

Water is an incredible substance. It has the power to sustain life, as well as being a key factor in some of the most dangerous and deadly weather conditions known to us. There are about seven billion people on earth and it is estimated there will be eleven billion by the end of this century—and by the end of this century the pressure on water supplies and the need for clean drinking water will increase exponentially. Dr. Bill Davies is a distinguished professor at the Lancaster Environment Centre at Lancaster University in England. Dr. Davies is a plant biologist and focuses on the effects of atmospheric and edaphic (soil-influenced) environments on plants. He states, most aptly, "Water is the new oil. People will be scrambling for water." And when people have to scramble for basic items, bad things happen!

NINE

La Réunion

The most beautiful thing we can experience is the mysterious. It is the source of all true art and science.

—Albert Einstein

Twenty-nine hours is a long time to get anywhere, but that's how long it took to get to La Réunion from my home in Lake Tahoe. La Réunion is French. But don't be fooled, though it is French, the island of La Réunion is nowhere near the country of France. It is more than 5,800 miles south southeast of Paris! The island of La Réunion is located in the Indian Ocean east of Madagascar, in the southern hemisphere, lying almost directly on the Tropic of Capricorn. The island is thirty-nine miles long and twenty-eight miles wide, and today has a population of 840,000. It is quite mountainous, with the Piton des Neiges volcano (extinct) rising to 10,070 feet above sea level, near the center of the island. I was here on

La Réunion as the recipient of a Fulbright Senior Specialist award. The Fulbright Specialists Program's goal is to link U.S. professionals (PhD required, plus a minimum of five years experience post-PhD) with professionals at counterpart institutions overseas in order to facilitate international cooperation and an exchange of ideas across international borders. I would spend three weeks on the island working and interacting with Dr. Jean-Luc Baray and the researchers and staff at the Laboratoire de Physique de l'Atmosphère, Le Centre National de la Recherche Scientifique (CNRS—the French National Center for Scientific Research), Université de La Réunion (University of La Reunion). I found out that the research being conducted at CNRS in La Réunion, despite being on this relatively tiny island that some might consider to be in the middle of nowhere, is surprisingly cutting edge.

Dr. Jean-Luc Baray was there to greet me at the St. Denis Gillot Airport, which lies on the north shore of the island of La Réunion. As soon as I walked outside of the airport I noticed how hot and sticky it was, unusual for the month of May in that part of the world, where it is usually cool and dry from May through November. But the stickiness was thanks to tropical cyclone (or *saison cyclonique*) Manou, which had literally just passed north of the island a few hours prior to my arrival and was now headed west toward Madagascar.

Very early on my first morning, I was surprised to be awakened by the Muslim call to prayer, the chant known as *adhan*, on loudspeakers being broadcast throughout the city of St. Denis. At that moment I realized that I had spent so much time researching the terrain, weather, and climate of the island and had completely neglected to get familiar with the culture of the island. I obliviously just assumed they would be mostly French, but the demographics of the island are actually extremely diverse. Some of the ethnic groups represented on the island include French, African, Chinese, Malagasy (people from Madagascar), Indian, and Pakistani. Although 86 percent of the population is Roman Catholic, Hinduism, Buddhism, and Islam are also practiced. The economy of the island relies heavily on financial assistance from France. Agricultural exports, especially sugar cane, dominate the island's economy, followed by tourism. However, poverty and joblessness are major issues on the island.

While doing research for this book, I came upon a wonderful publication, *The Pleasure Excursion: How One is Eager to Traverse the Regions of the World* (English translation). The original Arabic title of the book, considered to be one of the greatest works of medieval geography, is "Kitāb nuzhat al-mushtāq fī ikhtirāq al-āfāq" and was written by the Arab geographer Ash-Sharīf al-Idrīsī, sometimes referred to simply as Al Idrisi. He was born in Sabtah, Morocco, now a Spanish enclave in Morocco. As a "scholar of the earth," Al Idrisi studied under the patronage of Roger II of Sicily, the Norman king of Sicily, and wrote this book of geography in 1154, which included a world map he created for Roger II. His incredibly detailed maps even included climatic zones of the northern hemisphere.

What ties Al Idrisi to the tiny island of La Réunion is that he may have been the first to record the island on a map back in 1153, although this has not been fully verified. Nonetheless, we do know there are no indigenous peoples in La Réunion, as it was an uninhabited island when it was first discovered by Arab traders in the Middle Ages. In 1507, Portugal officially annexed it and ruled the island, although the French claimed it somewhere between 1638 and 1642.

Once I arrived on the island, my days became a whirlwind of activity. I was treated to lunches and dinners and discussed the scientific projects that were taking place there. I also talked about what I was currently working on, like the Perlan Project, which was of particular interest to the scientists on the island. Most of the scientists I was meeting were involved in the study of the tropospheric-stratospheric exchanges, and the Perlan Project dealt with these exact same exchanges while flying in those regions of the atmosphere. Sounds a little obscure, but these exchanges are an extremely important element of understanding weather and climate. The earth's atmosphere is composed of layers, the lowest one being where we live on the surface, the troposphere. Then come the stratosphere, mesosphere, thermosphere, and exosphere as one ascends.

As one goes higher in the troposphere, where we exist on a daily basis, the air temperature decreases, as most of us are keenly aware even on something as simple as a hike. As we ascend, there is a division between the troposphere and the next layer, the stratosphere, which extends up to 32 miles (50 kilometers) above sea level. This division is called the "tropopause."

This is where the air temperature no longer decreases but remains constant. Once you reach the stratosphere, however, the temperature then *increases* with altitude! The temperature continues to increase as we ascend in the stratosphere because of the ozone layer. This ozone layer protects the atmosphere below it (including the earth's surface) from exposure to ultraviolet radiation. The ozone layer absorbs this radiation, and thus increases heat. The earth's atmosphere bulges out at the equator and is thinner over the poles. Where the Perlan glider is flying, near the poles, the tropopause is at around 33,000 feet (10 kilometers) whereas near the equator it exists around 56,000 feet (17 kilometers).

To coincide with the increasing temperature in the stratosphere, there is vertical stability, since the temperature is now increasing with altitude. The lifetime of chemicals that exist in the stratosphere can be hundreds of years, whereas in the troposphere they get cleaned out by our weather systems and deposited on the earth's surface and in the oceans. It used to be believed that no weather existed in the stratosphere and that the troposphere and stratosphere never interacted. We now know this is not true. When large thunderstorms or very strong frontal systems occur, for example, these can cause airflow to inject itself up into the stratosphere and vice versa. This interaction between the stratosphere and troposphere can occur with strong mountain wave activity associated with the polar night jet, which, as we saw in an earlier chapter, was just the condition we were looking for to fly the Perlan to record altitude.

The stratosphere's role in the weather and climate that we experience down on the surface of the earth is still quite unknown, which is why we are trying to send gliders and other aircraft up into the stratosphere to further explore it. It is also why the scientists on La Réunion are so focused on researching the exchanges between these two layers of our atmosphere. In addition we are trying to better understand the effects of these exchanges on weather and climate, and what mechanisms are required for better understanding of coupling the stratosphere to the troposphere to better represent them in current weather and climate models. The role of water vapor and its transport into the stratosphere, a key part in radiation and the chemistry of the models, is quite uncertain. Atmospheric gravity waves play a large role in climate by causing mass exchanges between the troposphere and the stratosphere, but they are still quite crudely parameterized (input) into weather models today. The lower stratosphere wind and temperature perturbations (disturbances) may also impact tropospheric breaking wave activity. This can potentially have a serious impact on such things as aircraft flying at these lower stratospheric levels, as the strong temperature gradients may flummox air data systems installed on the aircraft, causing them to respond incorrectly to the conditions.

The "ozone layer" is a term many people may already be familiar with, thanks to the large amount of press devoted to the ozone hole over the South Pole and the accompanying concerns relating to human health. But in truth, there is "good" ozone and there is "bad" ozone. The U.S. Environmental Protection Agency (EPA) has a saying: "Ozone: Good up high, bad nearby." Ozone near the surface of the earth (in the troposphere) is produced by a combination of pollutants, and the global climate impact of this ozone is small. The EPA states that this "low" ozone is the main ingredient of urban smog. It is harmful to breathe and it also damages cells in humans, animals, and in vegetation.

Ozone in the stratosphere—the ozone that is "up high"—is the "good" ozone, as it protects life on Earth from the sun's harmful ultraviolet (UV) radiation, specifically, UV-B radiation. Ozone is made up of three oxygen atoms and this thus written as O_3. Ozone is produced naturally in the stratosphere when ultraviolet light (UV-C) from the sun splits molecular oxygen (O_2) into two separate oxygen atoms and these highly reactive oxygen atoms then react with more oxygen molecules (O_2) to form O_3 (ozone). In a similar way, ozone is also destroyed by solar radiation. Ozone is unstable and will combine easily with other atoms to make something new, depleting the good ozone that the Earth needs.

Most of the stratospheric ozone is produced in the tropics but, conversely, concentrations of ozone are not particularly high in the tropics due to the intense solar radiation, which contributes to a high rate of ozone loss. While ozone is produced in the tropics, it is transported by airflow from the tropics toward the poles, where it accumulates. Concentrations of ozone are actually highest in the cold subpolar regions (but not over the poles themselves). Concentrations of ozone are lowest at the poles, in their respective winters, when there isn't any sunlight to help form additional ozone along with the ozone being brought in from the equator.

Ozone absorbs the sun's highly energetic ultraviolet radiation and this energy is transformed into heat, thus leading to the warming of the stratosphere. The maximum ozone concentration occurs around 30 kilometers (a little over 98,000 feet) above the surface of the earth, where the Perlan glider will attempt to soar in 2016. At these altitudes, almost 99 percent of the mass of the atmosphere is located within the layer below 30 kilometers.

Now, back to the "ozone hole," which is not really a hole but a region of highly depleted ozone in the stratosphere over Antarctica. The depletion over the South Pole begins with the onset of the southern hemisphere spring, when the polar vortex is at its strongest. This circulation of the polar vortex jet stream exists from the upper troposphere all the way up into the upper stratosphere. The U.S. National Center for Environmental Prediction states that "the longevity of the ozone hole and the polar vortex are highly correlated." Ozone concentration is measured in Dobson Units. Historical records show that total column ozone levels of less than 220 Dobson Units were not observed prior to the year 1979. The value of 220 Dobson Units is used as the boundary representing ozone loss, or depicting the ozone hole region where values are 220 Dobson Units or less. Ozone varies as the polar vortex and its associated polar night jet rotate around the South Pole in relationship to the movement of the vortex. The northern hemisphere is getting very close to having its own ozone "hole," too, as its ozone concentrations decrease to near the 220 Dobson Unit level.

All of this is brought me to La Réunion. The location of this island, geographically and meteorologically speaking, is excellent for measuring and observing these stratospheric-tropospheric exchanges and their effects on the global tropospheric ozone budget. The island lies on the border of the stratospheric tropical reservoir, which means that the tropical convection along with the subtropical jet stream induce the stratosphere-troposphere exchange right over this island. Also associated with this convection is the practice of biomass burning, which entails the burning of vegetation, a common practice on the African continent as well as elsewhere around the globe, causing pollution and spewing particulate matter into the atmosphere that is then susceptible to getting into the stratosphere through these tropospheric-stratospheric exchanges.

The island's mountain facility is located at almost 7,100 feet above sea level and is on the northwest portion of the island. Thus it is free from the influence of local pollution sources and is able to accurately measure the study of large-scale transport of pollutants from the African continent, as well as their consequences on our atmosphere. Daily measurements of ozone using a LIDAR (Light Detection and Ranging) instrument are made from this facility on La Réunion.

Once I began speaking with the various scientists at CNRS at La Réunion and gave a presentation on what we plan on studying on our high-altitude Perlan Project flights, I realized that the scientists wanted us to bring the Perlan I glider to La Réunion to fly! *Well, that is interesting*, I thought to myself. I had to think this through before responding. We would have to use thermals, or rising warm air parcels, as opposed to mountain waves in order to soar here. Riding thermals for soaring is a common practice: instead of the glider riding straight up on mountain waves as if on an elevator, thermals require circling around and around the thermal itself, slowly gaining altitude with each gyration . . . and yes, as one might expect, this tends to induce vomiting in unsuspecting passengers (or our fearless pilots, in this case). Unfortunately, I have experienced this firsthand. But more important, in choosing La Réunion, where would the glider land if it ran into trouble? The island is tremendously mountainous and there are few safe places to "land out," which is when the glider cannot get enough lift to make it back to the departure airport. Glider pilots can land in large open fields—golf courses and the like—but this island doesn't have many of those, except in a few spots near the shoreline. I explained this to the scientists, and that was the end of the discussion of bringing the Perlan I glider to La Réunion. I discussed this with Steve Fossett later. "Really?" was all he said, followed by a chuckle. Perhaps he was disappointed. Perhaps he was relieved. Today's advances in unmanned aircraft and the availability of drones may present a viable solution for the scientists at La Réunion and future attempts at sending gliders up into the stratosphere in this fascinating atmospheric region.

During the last few days on my trip to the island, I finally had some time to myself. My now ex-husband, who was with me on the island, had amused himself during the previous few weeks doing who knows what all day after he dropped me off at the university each morning. During these periods of separation, I was able to revisit my painful personal life. I wanted out of what had become an unhappy marriage; I had wanted to leave earlier, but the timing was never "right," or so I surmised at the time while burying myself in my work and building my career. I recalled my parents discussing the fact that I traveled so much, perhaps thinking that it was a form of escape from problems at home. Also, I think the risk of hurting family—both his and mine—with a divorce was just too great for me to come to terms with for quite some time. I really loved and enjoyed his family, including his two grown sons. As with most things in life, my feelings of leaving the marriage came on slowly but continued like a steamroller—slowly but surely running things over and squishing them.

So what were we doing together on this island? We had planned for a long time that my husband would come along with me on this trip. I traveled 99 percent of the time by myself. If the trip didn't have something "fun, recreational, and exciting" my husband never came along. Coming along just to support me never seemed to be a consideration, and the knowledge of this continued to eat at me.

There were other issues that continued to gnaw at me here on La Réunion. His maniacal gym visits and steroid use were each increasing at an alarming rate. He would always insist that "these weren't steroids, but a precursor to them, and the body would only use what it needs," during our increasingly frequent fights. I didn't care how often he told me this. I did not agree. I witnessed firsthand the changes in him: the anger and rage he always had, but hid quite well, became oppressive; he began getting much bigger muscularly and was now even more obsessed with lifting weights at the gym. He was quite a bit older than me, so perhaps he felt some pressure to stay fit. But his competitive nature with me (and his eldest son) was becoming stifling, and it eventually wore me out. I suggested that he talk to a therapist but he angrily responded that they were quacks and only for people who were weak. Life is always a two-way street, but my lack of being supported and the private belittlement became too much to bear. I kept a game face in public, but inside I was aching. He also had a talent for wearing a jovial good-fellow mask in public. This Jekyll-and-Hyde trait was extremely aggravating. I often wondered if any of my friends saw through it. I'm not even sure my parents did.

My ex had even become obsessed with our BB gun and shooting at little animals. He didn't kill them; he just terrorized them by shooting near them, seeing how close he could come to hitting them. (As I look back on things, that's pretty much how he treated me as well.) I despised this new perverse bent of his. He would stand out on the upper deck on the second floor of our Lake Tahoe house and shoot down at little animals and birds. He was doing this more and more often. It was a couple of weeks prior to our leaving for my Fulbright in La Réunion when it finally happened. I knew it would. He had hit and gravely wounded a baby bunny. It was suffering so much that he walked down into the yard and crushed it with a large rock. He seemed upset with himself, but I was unable to feel sorry for him. The only feelings I had left for him were pity, sorrow, and anger over such a useless killing. I worried too what this behavior portended for my well-being and safety.

It was our final few days on the island of La Réunion and what should have been a wonderful time was dreadful. I announced that I was leaving him and wanted a divorce. We were "trapped" on the island for a few more days, and what better time to have this final reckoning without any distractions. We slept in separate queen beds in the beachside hotel and I kept the large hunting knife I had packed in my bag right next to me all night long. It was one of those knives with a serrated portion on the top of the blade. I had packed it in my checked bag for protection from strangers, the strangest of which was now my soon-to-be ex. He became more erratic and talked about killing himself, but I knew he was too egotistical to do that. It was a ploy to make me feel guilty, to try to manipulate me into staying. I wasn't even sure that he might not attempt to kill me "by accident" in a fit of rage.

As one might imagine, I didn't get much sleep those final few nights and was relieved to step on that airplane for Paris. I was very anxious to get home and move out—which I did immediately once I got back. I treasured the time I had with the scientists at La Réunion but unfortunately the experience had a very dark subtext, marking the culmination of a very unhappy and fearful time in my life. I wonder if any of my new scientist acquaintances felt the unhappiness and terror I was feeling. The atmospheric ups and downs over the island we were all fascinated by became a discomfiting metaphor for what was taking place in my personal life. I hoped it was well behind me as I flew away from the island, and, hopefully, into a new beginning.

TEN

The Wind Is from the Left

Mind like parachute—only function when open.
—Charlie Chan
Charlie Chan at the Circus (1936)

Now married to my loving and incredibly supportive husband, Alan Austin, I was living on the south shore of Lake Tahoe, Nevada, in 2006, with our beautiful son, Evan. One day as I was reflecting over the past couple of decades, a person who had great influence on my young student life, Dr. Devrie Intriligator, popped into my head. She was my brother Patrick's best friend's mother and president of the Carmel Research Center. I worked for her part of the time I was an undergraduate student at UCLA. She is a space physicist and still is one of my greatest role models. Part of her research involved studying solar wind. Dr. Intriligator received data from the *Pioneer 10* and *11* satellites and analyzed the solar wind. As the satellites travelled farther and farther away from Earth, she learned more about the solar wind, and is credited with many discoveries. I have listed some of them later in this chapter.

Solar wind is a stream of plasma, mostly electrons and protons, emanating from the sun. The Earth's magnetic field protects it from bombardment by these charged particles by deflecting most of them back out into space. Some of the high-energy particles make into the polar regions of the earth's atmosphere, causing the Earth's magnetic field to change shape and sometimes producing problems with electrical equipment on Earth. While I was working for Dr. Intriligator, she told me how male-dominated her field was back in the early 1980s, although it probably still holds true today. She told me about a scientist that didn't agree with her predictions about the solar wind and the *Pioneer* satellite missions. Devrie Intriligator predicted that as the first satellite passed Jupiter and began its trek into the outer solar system, the solar wind would not end there. Rather, she hypothesized that solar wind extended out far past Pluto. This other scientist did not agree with her at all. He was emphatic that the solar wind would end at the outer solar system. She did not see any evidence as to why this would be the case, and they argued about this fact . . . until the day of reckoning. The day Devrie analyzed the new data from *Pioneer* that showed the solar wind did not end at the outer solar system but continued right on as she had thought, she was rightly excited. The scientist with the dissenting hypothesis never spoke to her again.

Over the course of her career researching the weather in space (yes, there is weather in space—including the solar wind, magnetic storms, and turbulence), Dr. Intriligator became the first to discover a bevy of amazing new things. Here are just a few, which illustrate what an amazing impact this woman has made and continues to make on the field of space physics.

SOLAR WIND	First to measure in-situ space plasma density fluctuations
	First to measure the scale size of plasma turbulence in the interplanetary medium (solar wind), finding it was more than ten thousand times larger than the previously accepted size

	First to measure plasma turbulence in the asteroid belt First to identify interstellar hydrogen ions beyond the orbit of Jupiter First to recognize that the solar wind stream interface is a structural boundary to energetic particles in co-rotating interaction regions
VENUS	First to measure oxygen scavenged from the Venus atmosphere in the solar wind First to measure the hemispheric asymmetry in these oxygen ions
JUPITER	First to identify the Io plasma torus (a cloud of plasma that encircles the planet) at Jupiter in the *Pioneer* data First to identify the Europa plasma torus at Jupiter
EARTH	First to measure the Earth's extended magnetic tail beyond the orbit of Mars
COSMIC RAYS	First to measure the absolute flux of cosmic ray neutrons
INSTRUMENTS	First to design and implement a laboratory astrophysical plasma facility Inventor of cosmic ray detectors, space plasma detectors, and instrumentation First U.S. scientist invited to place an experiment on a Soviet spacecraft

As much as I revered and admired Devrie, I was not interested in pursuing a career in space physics and space weather, although she was an important role model for me as to what a career in science could look like.

My interests were closer to home, the Earth's surface, the troposphere and the stratosphere. But this was an important decision for me, as it set me down a path that led from UCLA to the University of Nevada, Reno (UNR) and the Desert Research Institute's graduate program, where I could get my fill of our weather here on Earth. I eventually became a professor at Sierra Nevada College and taught there for several years before founding WeatherExtreme Ltd.

The time I spent working for Dr. Devrie Intriligator opened my eyes to two things: women make amazing scientists, and yet they are not immune to experiencing sometimes uncomfortable or even unethical treatment from male colleagues.

It was a spring afternoon and at the time my husband, Alan, was flying corporate jets and helicopters for a living. He was nearing his destination, the Sea Ranch Airport on the northern California coast—a tricky non-tower airport to fly in and out of, as there is no one staffing the airport nor is there a weather station that reports on the conditions there. (The final approach path also required going directly over a Buddhist temple, which must have caused considerable havoc with the monks as they tried to meditate.) Prior to landing, his usual procedure was to call or have the other pilot call the guy who owned the deli/coffee shop at the end of the runway and get a weather update. Not terribly sophisticated, but Alan and other pilots had little choice. The owner, for whom this is routine, answered the phone and immediately said, "Hang on,"—he knew right away what the pilots wanted—which meant he was either going to stick his head outside or, if he was busy, just look out the window to do his weather report. A few moments passed and he came back. "Well, it's kinda cloudy, but it's kinda clear." "What about the wind?" the other pilot asked. "The wind is from the left," the deli owner reported, after looking at a nearby flag.

What does that even mean? You might wonder, as did the students to whom I told this story in my forensic meteorology class at the University of Nevada, Reno. I explained to the class that what this "wind is from the left" story illustrates is the importance of *proper* weather information for pilots.

The deli owner's description of the weather, though to him quite accurate, was little, if any, help at all to a pilot or a meteorologist. As scientists we need to use terms that we all understand so that we can communicate effectively. "From the left" means nothing to a pilot, as he has no idea what your left is. What he needs to hear instead is the azimuth direction from which the wind is coming and, hopefully, its speed as well.

Having been a professor at Sierra Nevada College, a four-year private college at Lake Tahoe, Nevada, and the University of Nevada, Reno, I had many capable students in my classes. The majority of my classes were comprised mostly of men, although there were a handful of women, especially in the required courses. I taught everything from snowmaking to calculus to physics to environmental science. One of my goals in life is to encourage more women to pursue careers in science and math through teaching and by example, by showing others that a woman with a PhD in such an obscure field as mine, atmospheric physics, can make it in today's world, hopefully just as I was influenced by the female science pioneers that preceded me. Though there were quite a few young women in the required classes I taught, unfortunately, the fact is that most of them would not continue in the sciences as a career. Statistics show that women receive only 19 percent of the U.S. physics doctorate degrees awarded each year, and fewer women are entering the fields of science and math. Studies have found that only 10 percent of faculty hired in the field of mechanical engineering are women. Women comprise almost 57 percent of the U.S. work force; yet they are a meager 8.5 percent of engineers and 12 percent of physics faculty.

What is the problem? I am convinced that part of it is the fear of mathematics being too difficult and male-dominated, along with the fact that girls are not always encouraged or told that they can (and do) excel at math, starting at the elementary school level. My biggest gripe about our society's attitude toward math, which extends beyond gender bias, is that is it taught in a manner that is incredibly boring. The focus is on rote learning. Students are not shown the incredible force of mathematics when applied to the real world. Real-world applications are the crux of understanding mathematics and why it is so important. Yes, I may make enemies with pure theoretical mathematicians—theoretical mathematics has its own place, but most of us need to see math applied to everyday life in order to embrace it. This is especially true for young people and children, in whom the love of math should first blossom. The very first math class that I really "got" and actually enjoyed was geometry, because I saw a direct application to the world I live in. This may sound crazy, but if we began to teach calculus—or at least calculus-like thinking—at a much earlier age, I believe that more women would go into the sciences. Why? Because calculus mirrors the language of science and it is much less about obtaining a single answer to a particular problem but a range of possible solutions. *Wow, just like the real world.* Calculus is basically the study of change. It is highly conceptual and I think more people, and especially more girls and women, would connect with this problem-solving approach. The social issues are even more complex, but I think that fostering the love of math at a young age is a key step in overcoming hurdles (institutional and social) that women in math and the hard sciences face later on in their prospective careers.

It was Friday many years ago, and Merton Burkhard had the projector all set up for our afternoon movie on physics. Mr. Burkhard, my physics teacher at Santa Monica High School (affectionately called "Samohi") turned on our weekly Friday physics flick. It was a black-and-white film, as usual, and the sound of the sixteen-millimeter Bell & Howell projector was almost as loud as the sound of the narrator in the film itself. Our class took place just after lunch and I have to admit sometimes the drone of the projector was just too much, and even I would have to try hard to keep my eyes open. Mr. Burkhard would walk the aisles between the desks so we couldn't sleep, but it was so tempting at times. Fortunately, not all the films made me sleepy. In fact, that day's movie would turn out to be one of my lifelong favorites, *The 1940 Collapse of the Tacoma Narrows Bridge.* Looking back on it now, it seems so obvious: it was the combination of physics and atmospheric science!

The Tacoma Narrows Bridge spanned the Tacoma Narrows of Puget Sound and connected the city of Tacoma to the Kitsap Peninsula, which lies west of Seattle. Though the concept of the bridge was kicked around for years, the construction on the bridge finally began in 1938. It was opened on July 1, 1940. Why a bridge to the Kitsap Peninsula? Military reasons. The U.S. Congress authorized six strategic areas of the country for the establishment of airfields, and Tacoma Field, owned by the city at the time, was named as the airfield that would service the Pacific Northwest. With the threat of war looming, the Roosevelt administration was funding more and more public works projects. On May 5, 1938, the City of Tacoma transferred ownership of Tacoma Field to the U.S. government, and it became McChord Field. This was just two months after Nazi Germany took over Austria. The bridge connected McChord Field and Fort Lewis south of Tacoma to the navy shipyard in Bremerton, the largest city on the Kitsap Peninsula. The cost estimate for the project was eleven million dollars, which would be 180 million dollars today. But in the end, only 6.4 million dollars were allocated, and the lowest bidder on the project got the contract at just under 5.6 million dollars. The lowest bidder was Leon Moisseiff, the designer of the Golden Gate Bridge, the famous suspension bridge completed in May of 1937. The cost to build the Golden Gate Bridge was 27 million dollars, almost five times the 5.6 million dollars available to build the Tacoma Narrows Bridge. In order to stick to the budget, the bridge had to be lighter and narrower than any other bridge the company had built, and five times cheaper.

The Tacoma Narrows Bridge was to be a suspension bridge just like the Golden Gate Bridge. Suspension bridges are designed so that the weight of the deck or roadway is supported by cables that are anchored at either end, and also are usually connected to raised towers at intervals along the bridge. Interestingly, all suspension bridges can "sing." Similar to guitar strings that vibrate and create tones, the bridge cables create various tones; however, most of the time these sounds are not in the range audible to the human ear. Bridges are designed to be able to withstand deflections from atmospheric and earth-generated disturbances; this is especially true of suspension bridges. The Golden Gate Bridge was designed for a deflection upward of 5.8 feet and downward of 10.8 feet. Its transverse (sideways) deflection design is 27.7 feet at the midspan, and it can withstand winds of 100 miles per hour. In contrast to the much heavier and substantial Golden Gate, the Tacoma Narrows Bridge was designed for a 20-foot maximum deflection (or sideways movement of the deck) and supposedly could withstand up to 120-mile-per-hour winds.

In fact, designer Leon Moisseiff is the one who introduced deflection theory in bridge design. Deflection theory describes how a roadway flexes with the wind, thus greatly reducing the stress by transmitting the forces from the wind through the suspension cables and into the bridge towers. Many forces impact suspension bridges. There are "live loads," or downward deflection forces, which are caused by the cars, trucks, and people that pass over bridges; there are "dead loads," or upward deflections caused by the forces from the form and the weight of the structure itself; and then there are forces of nature, which may be both vertical and horizontal (or transverse) forces, such as wind. Wind causes both static and dynamic loads, as the wind pushing along the side of a bridge causes it to push the bridge sideways. But this then gives rise to vertical motions in the bridge itself, which can create oscillations of the bridge in any direction. If a suspension bridge is not properly designed, these "transverse wind loads" can cause the bridge to go into a complex mechanism called "torsional flutter" (sometimes called "aeroelastic flutter"), which means that the bridge will begin to self-destruct.

However, wind alone is not sufficient to generate flutter in a bridge. The bridge design itself plays a prominent role in generating the flutter. Most people have experienced some sort of flutter or self-induced harmonic vibration in their life at some point. Riding a skateboard down a steep hill causes "speed wobbles." At some point most surfers, skiers, or aircraft pilots experience some form of flutter. It is not good, and we all try to get out of it before disaster strikes and we crash.

It was July 1, 1940, and the weather was beautiful. The Tacoma Narrows Bridge was officially opened for business. It was the third largest suspension bridge in the world and spanned 2,800 feet (the largest was still the Golden Gate Bridge, and the second was the George Washington Bridge). The total length was just under six thousand feet. It was impressive. But it had far less steel in it than the Golden Gate did. Remember, it had to be built on a drastically reduced budget. In fact, while the bridge was still under construction, one worker noticed that it "bounced." Then other workers began to notice how the bridge undulated like waves in water during certain periods, and they nicknamed her "Galloping Gertie."

This should have been a clue to the designers that something was severely awry, but during this time period, as David P. Billington, a bridge historian, claims, bridge designers had a blind spot with regard to this bridge, namely the fact that they didn't realize that wind created vertical movement of the bridge! They thought the wind could only cause horizontal movement. Although there had been examples of wind-induced bridge collapses in the past—including the collapse of the Falls View Suspension Bridge over Niagara Falls back in 1889, in which the bridge rose and fell by 20 feet and twisted up to 45 degrees—these bridge disasters had been inaccurately attributed to failures in the design or structure of the bridge, never to the wind itself. Compounding this was the fact that aerodynamic forces were still poorly understood in the late 1930s. That would soon change.

On the morning of the seventh of November in 1940, just four months after it opened, the Tacoma Narrows Bridge collapsed. The sustained wind gusts in the region were approximately 42 miles per hour. "Galloping Gertie" responded by beginning to oscillate. A bridge engineer noticed that the oscillations were getting worse, so he contacted Barney Elliott, the owner of a local camera store, to come quickly and take some footage as the bridge was in trouble. Barney and his friend, T. Harbine Monroe, scrambled to get their gear assembled. Upon arriving at the bridge, they set up their cameras on the bridge itself and on the shoreline. Four different Bell & Howell sixteen-millimeter film cameras were filming the events with full-color Kodachrome film. Paramount Studios eventually purchased the footage and converted it to thirty-five millimeter *and* converted it into black-and-white for use in movie theater newsreels around the world.

The bridge began twisting and undulating like it was made of jelly. People abandoned their cars on the bridge and ran to get off it. Mr. Elliott put down his camera, helped the last person reach safety, and then filmed Gertie's final moments. Elliott recalled in a later interview, "I watched it go down through my viewfinder, and then I ran like hell." There were no human casualties, but a dog trapped in a car was killed, as he was too frightened to leave the vehicle. Elliott and Monroe's film became very valuable for scientists and engineers in reconstructing what went wrong.

After extensive review, it was determined that the Tacoma Narrows Bridge collapsed due to torsional flutter. Though Moisseiff had designed or assisted in the design of many bridges, including the Manhattan Bridge over the East River in New York, the Benjamin Franklin Bridge across the Delaware River, and the Golden Gate, the collapse of the Tacoma Narrows Bridge effectively ended Leon Moisseiff's career.

Mr. Burkhard explained to our Samohi physics class what went wrong with the bridge. He went into great detail about the collapse of the bridge and the physics equations behind it. He even discussed the vortex shedding caused by the wind interaction on the deck of the bridge, exacerbating the twisting of the deck. I was not used to someone getting so excited about *physics*, and in this case the atmosphere was involved, too. It was great!

After teaching physics all day at Samohi, Mr. Burkhard would sometimes teach physics at UCLA in the evenings. He gave us Samohi students extra credit if we attended lectures at any of the local colleges or universities and encouraged us to think beyond high school physics. He took such pleasure and care in his teaching. He actually cared that we learned something. As I look back, I realize that Mr. Burkhard made me a better person. I was hooked and I didn't even know it yet.

A few years later, I sat on the top of the stairs in front of UCLA's Royce Hall, the one that overlooks Wilson Plaza, reading the complete college catalog course descriptions one by one. Something caught my eye: courses about clouds, the ocean, weather, and climate. They had names like: "Air and Water Pollution," "Cloud Physics," "Operational Meteorology," and "Ocean Science." *Why weren't these things taught in high school?* I thought to myself. *If they had been, I would have known what I wanted to do with my career, and my life, long ago.* Today schools regularly have entire course sections about the weather and climate, but back when I was in junior high and high school, other than the standards like the hydrologic cycle, the subjects of weather and climate were limited to a single chapter in a book, at best. I stood up, walked across campus to the Math Sciences Building, rode the elevator up a couple of floors, walked right into the main office of the UCLA Atmospheric Sciences Department, and introduced myself.

"Let me see if the chairman of the department is in his office," the lady at the front desk stated as she peeked her head into his office, which was just around the corner. Dr. George L. Siscoe, the department chair, was in. She asked him if I could come in and speak with him. He graciously welcomed me into his office.

"Welcome to the Atmospheric Sciences Department," he said, smiling, as he offered me a chair. Dr. Siscoe was just taking over as the department chair from Dr. Hans Pruppacher, who, in my book, is the father of cloud physics. Dr. Siscoe explained to me that this was the very first year UCLA offered an undergraduate degree, as for years the department only offered master of science (M.S.) and doctor of philosophy (PhD) degrees.

"I hear you're interested in our department," continued Dr. Siscoe. "How did you find out about us?" he said with anticipation.

"I read about your program in the catalog and realized that this is exactly what I want to major in," I replied. He then explained the program to me and revealed that they were looking for more undergraduates in their new major and that he was pleased to have me enter the program. I was reminded of the saying that "it is better to be lucky than good," as I felt like the luckiest person on the planet that day. I was accepted into the atmospheric sciences undergraduate program at UCLA, and I was also unwittingly about to plunge headfirst into what was then (and perhaps now) mostly considered "a man's world." The world of science. From my early beginnings in the sciences to now I have found that, in my opinion, women tend to bring a more all-encompassing view to problem solving. In addition, women are nurturing and sustain both themselves and others through adversity, two important qualities for succeeding in life. But women are highly underrepresented in the sciences. This must change. In order to solve the atmospheric and, thus, environmental problems we are facing, including the complex issues involved with climate change, women must play a larger role.

Though I do support and encourage women in the sciences, I do not think that the success of women in science has to come at the expense of men, or that there should be special treatment. Equality in all things. I do not believe in labels that pigeonhole people. However, when I was in high school, I was tremendously influenced by the book *Annapurna: A Woman's Place* by Arlene Blum. In fact, I believe that this book should be required reading for all high school girls. It is a true story of a group of women, a couple of whom became the first Americans (and women) to climb Annapurna, the Nepalese 26,545-foot mountain, the world's tenth highest peak. But the story is less about climbing and more about what women can accomplish and triumph over during their lifetime. Regarding my situation, I actually think that it was better for me to be one of only a handful of women in my area of study, as it increased my drive to succeed. Over the years I have noticed a large increase in the number of women entering the atmospheric sciences, and it makes me happy to see a new generation taking on this fascinating field. I welcome competition, and competition from both genders. What has accounted for this shift? Part of it may be due to the tremendous popularity of the weather sciences now, but, regardless, it is a welcome change. When I was a student at UCLA and I would tell friends of my parents about my major, they would often ask, "Are you going to be a TV weather girl?" No offense was intended, but it would be akin to asking a female aerospace engineering student if she was planning to be a stewardess, instead of an aeronautical engineer or a pilot.

Today, the atmospheric sciences are hugely popular with both men and women and the media has helped spread the word with stories on El Niño/La Niña, global climate change, air pollution, and extreme weather events. Throughout my career I rarely felt belittled or deprecated being a female atmospheric physicist, although there are plenty of instances of it happening to others. However, I knew early on that I must achieve my PhD in the field: being called "Dr." as opposed to "Ms." makes a world of difference when one wants to be taken seriously.

When I started pursuing my doctoral degree and began working at the Desert Research Institute, my first official day there drove the point home that there was quite a gender imbalance in the field.

"Hello, you must be the new secretary," the middle-aged lady behind the desk stated. "I will show you to your desk," she added without allowing me to answer.

As she began to walk from behind her desk, I blurted out, "No, actually I am a new graduate student here."

"Oh, I am so sorry. I just assumed . . ." she trailed off as she was in thought. "Really? Okay. Let me take you to the director of the graduate students office," she added as she led me down the hall to Dr. John Hallett's office. I found out later why she was so befuddled, I was one of only two women in the graduate department at the time, and I was the only American woman. In fact, I was one of only a handful of Americans in the entire department.

It was years later, when I had my PhD in atmospheric physics, that my gender would be used against me in an unlikely scenario. An attorney in Carson City, Nevada, retained me on a slip-and-fall ice case. The case was fairly straightforward and involved me testifying about the weather conditions including precipitation, temperatures, and sun position, which I did. As with most cases, there were other factors besides the weather involved. I was on the stand ready to testify. The jury was not in the courtroom yet and the attorneys were going back and forth with the judge about what I could or could not testify about. During this time the judge made a comment that was something along the lines of me being a "beautiful expert witness that the jury would like to hear from." I took it as a compliment and thought nothing else of it. The jury was not even present at the time. During a break in my testimony, while I was meeting with the attorney who had retained me, he proclaimed that since I had taken such offense to the judge's comments, he was going to ask for a retrial. Obviously the trial was not going as he had hoped.

I was shocked. I responded, "I did not take offense at all to the judge's comments," putting it to rest, I thought.

A couple of years later, I received a call from the same attorney. After he refreshed my memory about who he was and the case, he started right back in, stating, "I know you took such offense at the judge's comments to you while you were on the stand that I want to write a motion to have it retried."

I took a beat. "I remember what happened clearly and I was not at all offended by the judge's comments," I replied with conviction. And that basically ended the call. I never heard from him again. One's looks should not be an issue in the courtroom or the lab, with regard to either gender, yet it is an issue that is bound to occur more for women than for men. Though I was aware of this, I felt that continuing to protest the issue distracted from my real purpose there as an expert witness and was paradoxically making it *more* about physical appearance than about the evidence at hand. I was also keenly aware that no attorney would ever hire me because of my looks. They want to win cases and they do that by hiring the best experts they can find. I have come to be known as one of those, and that's why I'm called so often to offer my expertise and knowledge to aid them in their cases. I know clearly the reason I'm on the stand so often: I worked very hard to get there.

As a professor of atmospheric physics and an owner and founder of a woman-owned science-oriented company, I regularly deal with many of the issues that women face in the scientific and technical world. Statistics show that women receive only 19 percent of the U.S. physics doctoral degrees awarded each year, and fewer women than men are entering the fields of science and math. Some studies have found that only 10 percent of mechanical engineering faculty hired are women. Women comprise almost 57 percent of the U.S. workforce, yet only a meager 8.5 percent of engineers and 12 percent of physics faculty are women. It's worth noting as well that only 10 percent of bachelor's degrees and 6 percent of PhD's in physics are earned by Hispanics, African Americans, and Native Americans.

For me, being a wife and mother, and meeting the demands of running my own weather company is a daily balancing act that, as many other women know very well, takes time and effort (with, hopefully, a good deal of help from a spouse who is committed to sharing in the balance equally). Though all of my responsibilities bring me great pleasure, balancing the three is not always easy. Societal pressures do play a role. It seems that, in general, men do not struggle with these issues as much. Although this may not be true, it *feels* true to me. To achieve balance in life is an art—actually, balance in everything is an art. I find it interesting to see how other career-oriented women deal with these issues. What and where are the links? Understanding why so few women pursue careers in the sciences could make for numerous studies unto itself, and it's definitely something I would like to help change.

Dr. Zahara Hazari is an associate professor at Clemson University. She is working on reforming pedagogy in physics education, especially for females. In an interview about why women don't pursue physics careers, Dr. Hazari most aptly stated, "There's a general disempowerment in physics. Students perceive physics to be difficult. They have perceptions of it being close-ended. They perceive it as being unpleasant, and they perceive it to be masculine. It probably affects the level of gender participation." Dr. Hazari's research includes surveys of thousands of college students, and she discovered something very interesting. "The one factor we did find was that explicitly discussing underrepresentation had a significant positive effect on females' choice of a physical science career." I get this. It becomes personal to girls and women, and we begin to see that it is *our* choice whether or not we wish to change these patterns. It is up to *us*.

With the growing popularity of the environment and the atmosphere in education, things may change in terms of more women entering and *remaining* in the physical sciences. The fields of climate and weather do not have nearly as much of the "masculine-based" perception as straight physics. Yet in order to be adept in climate and weather, one must be proficient in mathematics, chemistry, and physics. This sometimes takes new students by surprise. After the first month of teaching the course Snow Science and Avalanche Control at Sierra Nevada College, during class one day, one of my (male) students blurted out, "This is my favorite class but it is so much harder than I thought it was going to be!"

"Yes, it is. It takes knowledge of math and science to understand the workings of our environment," I replied. And without missing a beat, I continued on with my lecture, with my high school teacher Mr. Burkhard not far from my thoughts.

ELEVEN

The Healing Power of Music and Women in Science

Where words fail, music heals.

—Hans Christian Andersen

"Dave Garroway left me a dollar in his will!" my father exclaimed. David Garroway was an American television personality and the founding host and anchor of NBC's *Today* show. It was actually quite an honor: Dave Garroway left a dollar in his will to everyone who made an important impact on his life. He was friends with my parents, and my father, Patrick Moody Williams, was one of his favorite composers. Garroway loved my dad's music, and near the end of his life he told my dad how much his music helped him through some difficult times in the hospital lifting his spirits and allowing him to forget where he was.

It was music that first brought my family out to southern California from my birthplace in New York City. I was four years old and my father got the chance to go from writing jingles in New York to writing music for film and television in Hollywood. It was great growing up in a house full of music; we all played various instruments. Most of the time my dad worked from home, writing music in a back office. My sister, brother, and I loved having him work from there. It was a great feeling coming home from school knowing someone was always there. You may even recognize some of my father's music, dear reader, as he wrote the scores for shows including *The Streets of San Francisco*, *The Mary Tyler Moore Show*, *Columbo*, *The Bob Newhart Show*, *The Days and Nights of Molly Dodd*, and *Lou Grant*. The wide variety of films he has scored are too numerous to list.

Music is found in every culture around the globe. Studies have shown that children who learn music in schools do better academically and excel particularly in math and science. On the SATs, music students' scores were 38 points higher on the verbal section and 21 points higher on the math section than the national average. Studies have found that music makers are 52 percent more likely to go on to college and other higher education than non–music makers. Music, art, and science have been integral pursuits of man for centuries upon centuries and I find it inexplicable that music and art are somehow delegated to the back row and have been basically shunned at our schools, colleges, and universities . . . a drastic mistake with drastic consequences! Dr. Lisa Wong discussed the fact that the fields of medical practitioners and musicians are complementary: for instance, both require the art of listening, as physicians need to listen to the patients they are treating, just as do musicians playing in an orchestra. Dr. Wong also mentioned how music can help our brains to have better aptitude for science and medicine. In addition, stroke victims sometimes regain their ability to speak through singing and dementia patients are known to become responsive instead of withdrawn when they hear music.

There is currently a big push in our schools to implement STEM (science, technology, engineering, and mathematics) programs or programs to promote STEM. In this chapter I thought it important to highlight the links between science, art, and music and why STEM cannot be singularly focused upon, to the exclusion of art and music. In this chapter, I'd like to focus on my life growing up in a house filled with music, playing various musical instruments, and the interplay between music and science. Music and the arts are a huge factor in the success of our young scientists, engineers, and mathematicians, and cannot pushed to the bottom of the ladder, so to speak, either financially or socially. Back in 1933, when he was playing in a musical trio on the S.S. *Deutschland* on his way to America, Albert Einstein stated, "If I were not a physicist, I would probably be a musician. I often think in my music. I live my daydreams in music. I see my life in terms of music." Note the musical propensities of Steve Martin, Woody Allen, former president Bill Clinton, Lester Holt, and many other people who love and make music in addition to their "day jobs"—Albert Einstein certainly isn't alone in expressing his passion for music.

There are many interesting and surprising facts about musicians. One is that music majors have the highest rate of admission into medical schools, followed by biochemistry majors. In the New York schools 90 percent of all students who participate in music go on to college.

In today's world of computers, tablets, cell phones, iPods, and the like, there is a tremendous emphasis on science, technology, and math, with music and art having been pushed to the back burner. I am a scientist and I think it's great that STEM is being implemented in schools and is being taken so seriously, but it should not be at the expense of the arts (including music). Dr. Frank Wilson, retired clinical professor of neurology at the University of California, San Francisco, School of Medicine and at Stanford University School of Medicine, researched the neurological effects of playing musical instruments. Brain scans show that playing music involves brain functions more fully (both left and right hemispheres of the brain) than any other activity studied. Dr. Wilson stated that music is an absolute necessity for the total development of the brain and the individual.

When my dad was twenty-four years old, he and my mom went to see William Steinberg conduct the Pittsburgh Symphony Orchestra with his parents, my "Mamaw" and "Bampaw." One of the associate conductors had just died, so Mr. Steinberg changed the program at the last minute to play American composer Samuel Barber's *Adagio for Strings* in honor of his life. My dad told me that it was, and still is, one of the most moving and beautiful pieces he ever heard. I know that this piece shaped his musical world, as I can usually tell my dad's music on the first couple of notes. I am never fooled, except when I hear the very beginning of *Adagio for Strings.* I always think it is my father's and then catch myself and realize what it is. I think it is the tremendous feeling that pervades Barber's piece and also all of my father's music . . . his music stimulates very profound feelings that one would not necessarily expect from meeting him casually—or even not so casually. My husband says that as soon as hears the first few bars of one of my dad's pieces, it's immediately recognizable as being composed by Patrick Williams. He also said dad's "music flows down the emotional crevices in your soul like a hot lava flow where tears are often born."

Charlie Rose interviewed my father on the show *CBS News Nightwatch.* Rose asked him about the piece he wrote, the *Theme for Earth Day.* This theme was for a two-hour TV program called *The Earth Day Special.* My dad's comment was, "I've never written anything for the planet before; it's a big responsibility." I wholeheartedly concur and am so very proud of this piece and what it represents. It combines the celebration of our Earth every year on April 22 and the wonderful world of music.

Regarding women and STEM, I think there has been a lot of progress in terms of more girls and women becoming interested in the sciences, technology, mathematics, and engineering. The issue still seems to be with them remaining in these fields over time. Many women still feel the "glass ceiling," the pressures of family versus career, and feeling the "odd man out" due to some arenas having the "old boys clubs." These are just a few of the issues women face during their careers and lifetimes. Another is: do we want to spend our lives working in a career that is highly male-dominated? This is what women who choose to go into fields such as mathematics and physics have to face. Overall, I think the impression is slowly fading away that the fields of mathematics, engineering, technology, and science are just for men. There are many fields of science that are filled with women (e.g., biology, geology, and zoology) and many other women tend to go into more "traditional" roles surrounding the sciences, such as teaching, nursing, or advising. Yet there are still very large gaps among the sexes in other areas, such as physics and mathematics.

More successful women in these fields will provide role models for future generations that will bring even more girls and women into these fields. These role models include women as professors, leaders of companies, and women in influential roles at high schools, colleges, and universities actively implementing STEM and women's involvement. It's not exactly a revelation that women do face quite a different set of challenges than men do, in terms of choosing—and staying—in a career. We must consider our blaringly loud biological clocks, whether we wish to remain at home with our children when they are young (admittedly, men face this issue, too), and if we wish to take on a job that requires shift work (e.g., weather forecasting) or tremendously long hours (running one's own business). I attribute much of the success of my company, WeatherExtreme, to the fact that Alan and I run it together. We can each take a load off of the other when needed. I understand that most couples do not share this situation, and so the importance of having a supportive partner cannot be understated—someone who is willing and able to help with childcare equally.

My own pregnancy was one for the record books. It went perfectly. I was forty years old and many of the articles I read all discussed the dangers of having children later in life, especially one's first child. The risk factors were all greater and much of the research focused on women forty and older. "Ugh," I remember thinking, "why did that have to be my exact age?" So I followed my doctor's instructions to the letter. And I was so thankful that I never felt sick during my entire pregnancy, as some women can really suffer! I even wrote a forensic report for a plane crash case and submitted it to the attorney the day before I went into labor.

Alan and I were so thrilled when our son, Evan, was born. My pregnancy had gone so smoothly that when the nurses asked us if we needed any help, we replied, "Oh, no, we have everything all set up in the nursery and are fine." Little did we know this was not going to be the case. Our son had to be delivered via an emergency C-section, as I had been in labor all day and our baby's heart rate was plummeting every time I was pushing. I had been pushing for over an hour. My doctor, the nurses, and Alan were coaching me through it, but finally our doctor, Dr. William Deschner, brought over the heart rate chart, at about one o'clock in the morning, and explained what was occurring. I immediately agreed to have the C-section, and within twenty minutes Evan was lying on my chest. He was perfect. As Dr. Deschner and the other surgeon pulled him out, Dr. Deschner exclaimed, "He's a bruiser." He was a beautiful, almost nine-pound strapping boy.

Soon after we brought him home, however, Evan began to have signs of the dreaded colic. I had heard of colic but had no idea what it really was. It is not an illness like a cold, but rather a situation in which a seemingly healthy baby cries all the time. Doctors think it might have to do with having "gassy tummies," but no one is truly sure! But to us it was like water torture, but instead of water, the torture device was severe lack of *sleep*. Everyone's sleep! By the second week of these colic events, Alan and I had to take sleeping shifts. He would watch Evan from nine P.M. until two A.M. I would set an alarm to awaken at two A.M. and take over until seven A.M. The concept of getting any sleep during my five-hour shift was out of the question. I might be lucky and get twenty minutes straight. Then Evan would begin crying. I would pick him up and carry him around, and he would stop crying. As soon as I just leaned back on a piece of furniture, not even to sit down but just to lean back on it, the crying would begin and I had to resort to walking around again.

Three weeks later in the midst of recovering from my surgery and dealing with extreme sleep deprivation, I received a call from an attorney I worked with, who said that I was needed to testify at a trial. I knew the trial was coming up but had no way to coordinate it with the birth of our child. The trial was regarding the June 1, 1999, crash of American Airlines flight 1420 in Little Rock, Arkansas, in which there were eleven fatalities and forty-five serious injuries of the 145 total on board. As I was to determine from my research, like many crashes, it was calamity born from a series of events that led to an inevitable catastrophic conclusion. The crew took the flight from Dallas, Texas, knowing that the weather was potentially an issue. To put it mildly, the weather was appalling (not a term I would use in court, of course). There were thunderstorms, lightning, and high winds—all the things that cause pilots to increase the expanse of gray hair at their temples. They also were up against the airline's ever-present "duty day" time limit issues which can often play a role in flight planning decisions. Duty times and flight times are spelled out in the Federal Aviation Administration regulations. They can be, and often are, further refined in the scheduling sections of the various union contracts with pilots. It was getting close to the end of their "duty day," and that can be a problem, both with the airline and the ever-vigilant FAA regulators, who can seriously hurt a pilot's career if he or she makes a misstep. The weather, especially the wind, at Little Rock was horrific. There were wind shear reports, lightning, and conditions were all changing very rapidly, making it very dangerous for the crew to accomplish a safe landing that night. They elected to give it a try and landed long (too far down the runway) and were, for a variety of reasons, not able to stop the aircraft before going off the end of the runway. Eleven people died, partly because of the poor decisions made by the crew, along with the other issues in the case cited above. If one considers all the factors involved in this crash, there can be some sense of understanding as to how it happened—not to say that it is "excusable." After many years of preparing for my role as an expert witness in forensic meteorology cases I have come to appreciate all the elements that come into play. These cases are, while tragic, always extremely complex which, of course, makes the often tragic outcomes even more heartrending. I bit the bullet as much as I could and was able to do my part on this one, even while not being able to attend the trial in person.

As time passed, we had, however, to do something about our colicky little bundle of joy. When Evan was two months old and we looked like we'd been on a drunken bender for months, we were finally able to find someone trustworthy to watch him during the day while we worked out of our home office. Going into separate offices during this time was completely out of the question. Between being so exhausted and my wanting to be in the same location as our new son, we converted the entire downstairs portion of our home into offices. Luckily, this worked out quite well for us and our employees, who did not telecommute, but were still able to staff the office in our absence.

While I have been fortunate in being able to make the choices I have made regarding family and career, another choice that sneaks its way into the discussion is in regard to names. When women get married, they must decide whether or not to take their husband's last name or keep their maiden name. Some choose to take the husband's last name and continue to use their maiden names professionally. It sounds simplistic, but this is a difficult decision that has ramifications the rest of your life. During my first marriage I took my then-husband's last name, Carter. I went from Elizabeth Jean Williams to Elizabeth Jean Carter. I decided to keep my paternal grandmother's name as my middle name. My mother, Catherine, went through the same process. But she chose her maiden name as her middle name, Catherine Greer Williams. Then, when she and my dad had me, their first of three children, they wanted to name me Jean Elizabeth Williams. But my mom realized that my initials would have been *JEW.* Even if we were Jewish, those initials might have made clothing monograms and other applications somewhat distracting, so instead they chose to name me Elizabeth Jean Williams.

Now, long after my divorce, I am happily remarried. In my second (and final!) marriage, to Alan, again I faced the name challenge. I chose Elizabeth Jean Austin. When I told my sister, Greer, of my decision, she said, "I love it, it sounds like the name of an author or a spy!" It was difficult to get people used to using my new name at first, but I have no regrets in choosing to use my husband's name. It fits me and my personality.

But the naming saga didn't end there. A year after we were married, Alan and I now had to choose a name for our son. We tossed all sorts of family names around. There were already so many Patricks, Christophers, Wilsons, and Williams (Bills), and simply Williams, that we decided to use Moody as his first name. We are not some hippies naming our son some crazy name like "Running Spring" or other LSD-induced moniker. This name goes way back in my family, as my dad's middle name, my grandfather, Wilson's, middle name, and my great-grandmother's last name. It is an English surname and is actually quite common in England; it is only us Americans who sometimes find it odd. His middle name would be Evan in honor of our friend who introduced us, Patrick Evans Bailey, Esq. Moody Evan Austin. That would be his name. But now Moody prefers to go by Evan due to being teased about his name by elementary school classmates, so Evan is what he goes by. What's in a name is a lot! Some women choose to keep their maiden names because they are worried about losing name recognition in their field, or about not being as easily recognizable for works they published before they were married. There is no right answer, and it is just one of many issues women face during their careers and lifetimes.

TWELVE

A Toxic Awakening

How did it get so late so soon?

—Dr. Seuss

What was happening to me? I was a teenager, and was confused about more than the usual changes teenagers face. If I tried to stand up, my ears rang, my vision became tunneled, and I instinctively dropped to the floor to avoid passing out. I spent the next two weeks in my parents' master bedroom lying in bed, sweating, freezing, sometimes delirious, and having nightmares when I did fall asleep. My mom took me to our family physician, Dr. John Opdyke, who tried to diagnose what was wrong with me. My parents met Dr. Opdyke when they were in college at Duke University, and John Opdyke was a medical student there. We knew his wife and kids, too, so he was truly our family physician. Day after day, Dr. Opdyke poked and prodded me, drew my blood, took saliva samples, and asked me all kinds of questions. The doc wanted to put me in the hospital but my mom refused. I found out later that my dad thought it was a good idea to put me in the hospital, too. I cannot imagine how terrified my parents were as my illness went undiagnosed day after day. I was completely unaware of things and just struggled to make it through each day. In retrospect, I was glad that I stayed at home, as I didn't realize how sick I really was, and just thought I would get better. If I had gone into the hospital, I might have panicked and in fear I might die.

When my fever got too high, my mom would put me in a cold shower to cool me down. My mom had someone help out during the day while she was busy taking care of me; my dad would pop in to see how I was doing, but I think he was scared because I was not getting better. I am not sure what my sister and brother were up to during this time, but they were probably worried and just trying to stay out of the way. Each day was the same: I slowly got out of bed and literally crawled down the hall, down our hardwood stairs, out the back door on my hands and knees and into our garage, where I would pull myself up into the car to be taken to the doctor. Dr. Opdyke would examine me and, as all the test results were negative, we had no new clues to go on. During the second week of my illness I started to realize that things were not what they seemed: my mom and I arrived at Dr. Opdyke's office, and he had two specialists there with him. The three doctors spent time poking and prodding me as I answered all sorts of questions—do you feel achy, does your head hurt, are you vomiting, does it hurt here—as they pushed on my abdomen. This went on for about twenty minutes. Then my mom took me back home where the cold showers, strange dreams, sweating, and chills continued, seemingly endlessly. Then I woke up one day, and I was better. I was very weak from being bedridden, but I felt as if I were nearing normal again.

I was fifteen years old and in the tenth grade, and had recovered from a mysterious illness that bewildered the doctors. I did not care that my illness was a mystery. I was just happy to be alive and healthy! My appetite was back and I got stronger every day. Three weeks after my ordeal with this strange illness began, Dr. Opdyke's diagnosis was that I had Kawasaki disease. What was that, I wondered. I soon learned that it is a disease that afflicts mostly children under the age of five and most commonly Asian boys. The disease is most prevalent in Japan and was named after Tomisaku Kawasaki, the Japanese researcher who first described the ailment in 1967. The disease attacks the coronary arteries and, if untreated, causes severe damage to the blood vessels that can turn into dangerous aneurysms later in life. My white blood cell count seemed to be the only thing that matched the symptoms of Kawasaki disease. I wasn't buying this diagnosis and neither was my mom. The detective in me was on alert that something was amiss. *I am going to get to the bottom of this*, I thought, *just like Miss Marple*. I read Agatha Christie's books cover to cover and particularly loved the ones with Miss Marple as the heroine, the unassuming, elderly solver of mysteries and crimes. One thing that would help me in solving this mystery was the fact that after I recovered, my hands and feet peeled. I noticed a little bit of skin beginning to lift up, and I pulled on it, and my skin pulled up in large pieces, similar to when my pet snakes shed their skin. It was so unusual, that had to be a primary clue, I thought. My mom and I began to gather medical reference books and news articles from the library. There was no such thing as Google at the time, so it was slow going as we pored through the books and articles, narrowing down the possibilities over the next couple of weeks. During this time my life was getting back to normal. I was back in school, enjoying my friends, and riding my beloved horse, Vandy.

As we researched, I learned that peeling of hands and feet is more common than I thought. It can be due to allergic reactions, infections, immune system disorders, and even cancer. We were narrowing down the suspects. My mom thought it might have been toxic shock. But I never used the brands of tampons that were the supposed culprits, I explained to her. I even told her that about six months earlier my aunt Katherine had warned me about the brands of superabsorbent tampons and girls getting toxic shock, so I had made sure to stay away from them. As our detective work continued, my menstrual cycle began. Thinking nothing of it, I used my regular Tampax tampon, and within thirty minutes of inserting the tampon I became dizzy, disoriented, and had tunnel vision when I stood up. *Oh my God*, I thought, *it is the tampon!* I rushed into the bathroom, trying to avoid fainting on the way, pulled out the tampon, and within ten minutes was feeling much better. My mom was right after all. It was toxic shock. I went back to the medical reference books and, yes, almost every sign and symptom matched mine. As I read more on the disease, the luckier I felt to have survived. Toxic shock syndrome is rare and life-threatening, and is caused by toxins produced by *Staphylococcus aureus* (staph) bacteria. The cause: superabsorbent tampons. Women and girls died of this, and it was becoming more newsworthy by the moment. Ever since the day I yanked that tampon out, I have not used another one ever again. I did not want to risk my life for this small monthly luxury. I was going back to stone-age, but safe, feminine pads. And that was that. The illness never came back. Though advances in medical technology, tampon manufacturing, and public awareness have reduced instances of toxic shock to a very low level in the past two decades, toxins are not something we can ignore, either in our bodies or the world around us.

"Our planet too is suffering from toxic shock, and we desperately need to do many things to save its life—NOW!" is how the headline *should* have read. By European Union air quality standards, of the 560 million city dwellers in China, only 1 percent of them breathe air that is considered safe. To drive the point tragically home, cancer is now the leading cause of death in China. Around the globe, premature deaths from both heart and lung disease have been linked to air pollution. Studies have linked low birth rates in Salt Lake City, Utah, to the poor air quality. Pollution levels have been reported to be increasing at an alarmingly rapid rate in the arctic, even appearing in food sources such as whale meat, which now contains pollution toxins. Trash and chemicals have been found in our waters all over the globe. Pollution is transported via air, ocean, underground water supplies, rivers, and into our food supply. Amazingly, some of this pollution is spreading, winding up in some of the most remote places on the planet. How does it get there?

I was discussing the issues of weather, climate, and pollution with my husband, Alan, and he made a very simple but astute statement: "People don't think of the air as a fluid and that's the problem." He is exactly right.

One of the very first courses an atmospheric sciences major takes is fluid dynamics (along with a multitude of subjects from chemistry to physics to geology to biology). Unlike some of the core sciences such as physics or chemistry, the atmosphere (as well as ocean) sciences require one to be a jack-of-all-trades, so to speak. But fluid dynamics, or the nature of fluids, their flows, turbidity, and mixing (or not mixing), is of the utmost importance in understanding the nature of the atmosphere, as the atmosphere is indeed a fluid. And as a fluid, it has waves, turbulence, flow, and thus it transfers the things in it—including toxins and pollutants—from one place to another.

Air pollution concentration and dispersion are determined in large part by meteorological conditions. How far a particular type of pollution travels and how concentrated it remains are dictated by the pollution type and the atmospheric conditions. For example, pollution can travel across continents and wind up in remote locations around the globe far from its source. Air pollution does not recognize political boundaries. Scientists are learning more about this by leaps and bounds every day. I remember when the term "acid rain" became a passionate topic in the 1970s and 1980s. Though the term was coined back in 1872 by Robert Angus Smith, a Scottish chemist, acid rain only became front-page news beginning in the 1970s, when various governments began efforts to reduce the emissions of harmful gases into the atmosphere. The two most notorious harmful gases in the governmental hit list were nitric and sulfuric acids, specifically nitrogen oxide (NOx) and sulfur dioxide (SO_2). There are others, too, such as ammonia and other volatile organic compounds. For you non–chemistry nerds, nitrogen oxide (commonly called "nox" by scientists) is a term that represents nitric oxide (NO), nitrogen dioxide (NO_2), and nitrous oxide (N_2O). NOx and sulfur dioxide gases are produced both naturally and through human endeavors.

These gases are emitted naturally from volcanoes, forest fires, and natural processes, such as the nitrogen cycle, including the breakdown of nitrogen in soils and the oceans. However, the *majority* of the emissions of these gases are from human activities. In fact, a whopping 99 percent of all sulfur dioxide emissions are from human actions. In the United States, two-thirds of the sulfur dioxide emitted into the atmosphere comes from the generation of electric power by the burning of fossil fuels (like coal). Auto emissions are another source of both sulfur dioxide and nitrogen oxides. Nitrogen oxides are also produced through a wide variety of industrial, and especially agricultural, processes. Nitrous oxide (N_2O), remember, is a member of the nox family, and is also a greenhouse gas. NOx are poisonous, highly reactive gases formed from the burning of fuel at high temperatures, such as emissions from automobiles, boats, power plants, cement kilns, and industrial boilers.

Back to acid rain. The term is actually somewhat misleading, as this acidic pollution actually falls in both dry and wet forms. The wet is rather obvious, as the chemicals interact with wet weather and fall to the ground when it rains, snows, sleets, or even with the onset of fog. The dry form is also quite common, although less obvious, and occurs when the chemicals become incorporated with dust particles, smoke, or aerosols, and fall to the ground. Aerosols are microscopic particles in the atmosphere. Thus, the more accurate term is "acid deposition" and not acid rain. But the term persists as a reference to both iterations today.

One place that was hit hard back in the 1970s and 1980s by acid deposition is the eastern United States. Entire forests were dying and ecosystems were suffering due to these acids falling into rivers, lakes, and towns. The states that were most affected were Illinois down to Tennessee, western North Carolina and up into the Ohio Valley region and into some of the northeast states. This is because these regions are situated east of many power plants, and the weather across the United States generally flows from west to east. This is obvious when watching the weather on TV as they show the jet stream or river of air up around 30,000 feet flowing across the country, sometimes with wiggles in it and other times fairly straight.

Some of my research over the years has been related to acid deposition. There was a link between my research on acid deposition and my childhood. As a child and teenager, I spent many of my summers with my grandparents, Jean and Wilson Williams, at their house in the mountains of North Carolina. My sister, brother, and our cousins all called Jean and Wilson "Mamaw" and "Bampaw". Mamaw is still going strong and is one hundred and one years old, but she no longer lives in the Mountain House, as we all called it. Their house was at Grandmother Lake, which is about one and a half miles southwest of Grandfather Mountain. Grandfather Mountain rises to almost 6,000 feet above sea level, which is high for a mountain in the eastern United States. Though many maps label the lake as Grandfather Lake, its real name is Grandmother Lake. We knew this because Hugh Morton lived just two doors down from my grandparents and explained the back story of the curious names. Back in 1952 Mr. Morton inherited much of the land, including Grandfather Mountain, from his grandfather. He knew that area of North Carolina like the back of his hand. We spent many wonderful days with Mr. Morton hiking, sailing, and listening to stories during cocktail hour in the evenings with him and his wife. My grandparents' house was right on the lake and there was a long, steep path down where they had a dock and small shed where they kept fishing and boating equipment. Coming from California, I was always surprised by the number of insects and creatures that lived around—and unfortunately sometimes in—their house. I saw bats living in the eaves above their deck outside, salamanders under almost any large rock I turned over, and lots of snakes. I was not afraid of snakes as we had them as pets at home, but I was not keen on the spiders or the centipedes.

Hugh Morton was a Renaissance man if there ever was one. He was actively involved in supporting and promoting the arts, tourism, libraries, education, and environmental issues. He actively battled the National Park Service to prevent "cut and fill" road building techniques from being used on the Blue Ridge Parkway; led the way to plant wildflowers along primary North Carolina highways; helped pass the "Ridge Law" in North Carolina, outlawing any new structures taller than thirty feet along the ridge line of the Blue Ridge Parkway; and his efforts led to eliminating "straight piping" sewage into mountain streams. No wonder my Mamaw and Bampaw always praised the great work that Mr. Morton accomplished. During my UCLA undergraduate years, Mamaw sent me some information about Mr. Morton's rally against air pollution and acid rain. This was another turning point in my life, and by the time I reached graduate school I was immersed in researching this toxic phenomenon. During my graduate school years, Hugh Morton photographed and produced *The Search for Clean Air*, narrated by Walter Cronkite and aired on PBS. This one-hour video analyzed the effects of acid deposition and air pollution in five different countries around the globe. The effects include damage to natural ecosystems, humans, and materials. Some examples of these damages include dying forests, more respiratory problems than ever before, and life forms vanishing from our streams and lakes, although that is just a small sampling. Just as important, the film offers suggestions—including conserving electricity, winterizing one's home, recycling, and cutting back on automobile usage—for reducing and preventing the pollution that causes acid rain. These suggestions are not to be taken lightly, as small changes can make a big difference, which reminds me of the quote from Christine Todd Whitman, former director of the U.S. Environmental Protection Agency: "Anyone who thinks they are too small to make a difference has never tried to fall asleep with a mosquito in the room."

Mamaw had just returned from a lecture that Mr. Morton hosted about how the forests of North Carolina and across the eastern U.S. were dying due to acid deposition, and she picked up the phone and called me from her North Carolina home.

"Liz, the forests in the eastern U.S. are in terrible shape. Mr. Morton has some amazing photographs showing the changes that have been occurring over the years. They believe that the acid rain is causing most of it."

"Mamaw, we are collecting air samples from the top of a mountain in Colorado that have pollution traced from China," I said as I held the phone to my ear, as I was now in graduate school, and continued, "and the East Coast of the U.S. has to deal with all of the coal-fired power plants in the Midwest and the East. The weather basically travels from west to east across the U.S., bringing the pollution with it."

"Mr. Morton also has photographs taken from Grandfather Mountain in the past and the present showing how the visibility in the air has deteriorated," she responded.

"I can believe it," I responded, "as there are sources of pollution all over the U.S. that potentially impact the East Coast. There are even sources people do not like to discuss. Cases like this were discovered at the institute where I am working."

"Really?" I had piqued her interest. "Like what?" Mamaw asked.

"It is a political issue, but there is much more pollution affecting the Grand Canyon and the visibility there than ever before"—which is a big deal, of course, as that is why people are drawn to it in the first place. "These air pollution scientists found that during certain weather conditions, one of the big polluters are the American Indians. This is obviously a sensitive subject as they operate under different regulations and laws regarding waste and pollution than everyone else in the U.S.," I told her.

"Well, remember the bald eagle pair Mr. Morton rescued?" she asked, in what seemed like an abrupt change in subject.

"I sure do. That was also a tricky subject but Mr. Morton handled it very well." I recalled how Mr. Morton rescued a pair of bald eagles (one female and one male) that had been shot. While nursing them back to health, he kept them on an island in Grandmother Lake. The island was just sixty yards from the shoreline of Mamaw and Bampaw's property. I was visiting one winter break in high school, which happened to be when Mr. Morton was caring for the eagles. The island was a safe haven for these incredible birds. It was about a fifteenth of an acre but it did the trick, as it was forested and had lots of bushes and trees. The eagles had many places to rest and feel secure. But Mr. Morton had to go out there once or twice a day to feed them.

Thinking back to that winter, I remembered when Mr. Morton phoned my grandparents and asked if I would like to help him feed the eagles.

"Of course!" I couldn't get the words out fast enough.

"Okay, meet me at your grandparents' house in fifteen minutes," he said.

I ran around the Mountain House locating my winter jacket, boots, and gloves. It was around 20 degrees F outside and cloudy. While throwing on my gear, I told Mamaw that I was on my way to the island to feed the eagles with Mr. Morton. Jean, being the Amazon she is, came with Mr. Morton and me down to the dock to see us off. Their trail down to the lake was quite steep so it zigzagged across the hillside. Not surprisingly, Mamaw was the first one down to the dock. Mr. Morton and I took my grandparents' small aluminum fishing boat out of the boathouse and put it in the water next to the dock. We loaded up our gear: most important, the food for the eagles. Mamaw saw us off and headed back up the hillside to the Mountain House.

We arrived a few minutes later at the island. We hopped out of the boat. Mr. Morton tied the rope extending from the front of the boat to a bush. Then he carried the bucket full of raw meat he bought from the local butcher over toward the center of this very small island. The bucket was full of long strips of raw meat; I cannot remember what kind of meat it was, but it was light colored, like chicken.

As Mr. Morton was on his way, I just happened to glance back toward Mamaw's house and saw that the boat had come untied and was floating away! Oh, God, I thought as I ran and quickly grabbed the very end of the rope that was now about three feet off shore and in the water. My arm all wet, I tied the rope securely to another bush and hurried to catch up with Mr. Morton. He was so focused on the eagles he never even noticed (and so I said nothing about almost being stranded on the island).

I began scanning the bushes for the eagles when Mr. Morton whispered, "There's the male," pointing over toward the edge of a small clearing where a portion of a tree trunk lay sideways on the ground. Then Mr. Morton spotted the female and quietly pointed her out. She was holding very still under a bush a few yards away from the male. I followed behind Mr. Morton as he slowly walked toward the fallen tree trunk. He began to speak to me in a low, non-whispering voice now.

"I always put their food on this part of the tree trunk," he said as we reached the trunk. The tree trunk had all sorts of bends and twists and turns to it so that the middle part of it was a few feet off of the ground. He began laying a few of the long strips of meat over various parts of the trunk. As he did, it began snowing, but he didn't even seem to notice.

While we were putting the strips of meat all over the elevated portion of the trunk, the male eagle suddenly semi-flew, as best he could after being shot, and landed on the end of the trunk not five feet from us. I had seen bald eagles soaring in the air and high up into the trees, but to be this close to one was incredible. He was huge! His beak was gigantic and his talons were so thick and long I couldn't believe it. Later I learned that an eagle's beak and talons grow continuously during its life and are only worn down through use. The wingspan on that bird must have been at least seven feet. As we continued laying the meat down, the male would take little hops along the tree, slowly working his way closer to the meat, and to us! I tried to act calm and just pretend that I was not interested in him, but that eagle was just amazing, and I couldn't help but just stare at him.

Mr. Morton and I finished laying the pieces one by one until we emptied the bucket. We slowly walked backward away from the tree trunk, and after just a couple of steps backward, the male was already eating a piece of the meat. At that point the female cautiously began walking out from under the bush where she had so quietly remained this entire time. The farther Mr. Morton and I walked away, the more steps the female took toward the food. Then apparently she no longer felt threatened, and she semi-flew, as best she could, up to where her mate was and began eating. I swear she seemed even larger than her mate.

On our way back to the "mainland" in our boat, I asked Mr. Morton all about the eagles. He told me that someone brought them to him to care for because of his reputation for rescuing and supporting animals. (There is a famous bear, Mildred, that he brought to Grandfather Mountain, and rehabbed back to health. Mildred lived out her years in the animal sanctuary created by Mr. Morton at Grandfather Mountain.) Mr. Morton continued, "But the sad part of the story that I don't tell many is that although these birds are protected under a federal act from being hunted, bought, or sold, these birds were shot by someone," he said.

"What happened to the people who shot them?" I asked.

"Nothing, because it was discovered that they were shot by some American Indians and there are some political issues there, because they can use their feathers and such for their religious ceremonies." He trailed off in thought. We were nearing the shore when he said, "Aren't they magnificent creatures." I could tell it wasn't a question but a reflective statement. I remained quiet.

The bald eagle shooting and the Grand Canyon pollution are just two small examples of what makes tackling pollution and climate change problems very difficult. Not only are there naysayers to the overwhelming scientific evidence of climate change, but also obstructive political ramifications which make the implementation of the necessary changes to do something about it difficult as well. Adding even another dimension to the problem is that some countries are now just becoming industrialized, "spreading their wings," and while doing so, they feel that they are being singled out in terms of having to cut back on their expansion and urbanization. All in all, it is, and promises to be for some considerable time in the future, a very thorny problem—and an increasingly critical one.

Years later, Mr. Morton's Grandfather Mountain became the first privately owned property in the world to receive recognition as an "International Biosphere Reserve." These reserves are recognized by UNESCO (the United Nations Educational, Scientific and Cultural Organization) and their purpose is to promote sustainable development based on local community support and sound science. The Grandfather Mountain Animal Habitats now have many animals, including deer, eagles, black bear, otters, and mountain lions. After Mr. Morton's death in 2006, his family sold more than 2,600 acres of land to the state of North Carolina, which is now North Carolina's thirty-fourth state park, Grandfather Mountain State Park, which is still going strong to this day.

It is ironic that all these years after surviving toxic shock, the disease that Dr. Opdyke mistakenly diagnosed me with, Kawasaki disease, is now linked to weather patterns. The spread of the disease to Japan and the West Coast of the United States is now linked to winds blowing from agricultural areas in northeast China. Apparently, scientists have known for years that the spread of Kawasaki disease had something to do with wind patterns, but for the first time those patterns have been tracked to a specific location, as discovered by Dr. Jane Burns. Dr. Burns is the director of the Kawasaki Disease Research Center at the University of California, San Diego, and Rady Children's Center Hospital in San Diego, California. I find it amazing that there is an entire research center devoted to this disease that I hadn't heard of since I was fifteen. There were 5,447 cases of this disease in the United States in 2009 and the incidence has been *increasing* throughout the years, in stark contrast to when I first heard of it, when it was very rare. Dr. Burns, Dr. John Ballester from the California Institute of Technology, Dr. Daniel Cayan of the Scripps Institute of Oceanography, and others from institutes in Japan and Spain analyzed climate data, air samples, and air flow back to 1977 and found a link between the peaks in outbreaks of Kawasaki disease, usually in the month of March, in the western United States and an agricultural area of northeastern China where cereal grain production occurs. The wind patterns bringing the disease to the United States are influenced by the El Niño/Southern Oscillation (ENSO) in which strong zonal winds (i.e., nearly parallel to lines of latitude) develop in the upper troposphere sweeping from Asia to the western north Pacific in the wintertime. But what exactly it is *in* the air that is causing the disease is not yet known. However, the air samples show high concentrations of *candida* fungus. *Candida* is a genus of yeast and is an organism responsible for most of the fungal infections in humans. But whether *candida* plays a role in Kawasaki disease is still unknown. Nevertheless, the link between this disease—caused by an unknown toxin—and the weather is proven and cannot be ignored.

In addition to Asian dust, there is African dust that is transported over portions of the globe. In fact, in addition to the outbreaks of Kawasaki disease in Asia, there are outbreaks in the Caribbean, Britain, and other countries affected by transportation of dust from Africa, where evidence that the pathogenic fungus *Aspergillus sydowii* is transported by these Africa (and Asian) dust storms.

With more than seven billion people living on our planet, the world is becoming a "smaller" place. By the year 2050, the population is expected to grow to nearly ten billion. There is and will be a tremendous need for safe food and safe drinking water around the globe. The pressures of food production, combined with deforestation to make land usable for agriculture and for housing and urbanization, are not only contributing to global climate change but also to the increasing impact of airborne diseases, which are spread through the transport of pathogens in the air. This spread is not just local, as from a sick person coughing into the air around us, but through the global transport of these pathogens through the atmosphere.

But there is some good news, as power plant emissions are increasingly regulated. The amount of acid deposition in the United States has decreased over the years, and though it is still an issue, the severity of it is diminishing. Mr. Morton spent years actively battling to clean up air pollution in North Carolina, much of which was drifting into the state from the Tennessee Valley Authority's thirteen coal-fired power plants; he would be pleased with the progress that we are making today with the push toward cleaner, safer energy. Just imagine how different our politicians and business leaders' thinking and planning would be if they lived for four hundred years and had to deal with the long-term results of their short term—often solely profit-driven—decisions!

Cleaning up the toxins in our world is not an option, it's an imperative. Our planet and are bodies are inextricably linked.

ABOVE: Hugh Morton and Mildred the bear at Grandfather Mountain, North Carolina. *Photo courtesy of the Grandfather Mountain Stewardship Foundation, grandfather.com.* BELOW: Jacob van Ruisdael, Dutch painter, oil on canvas painting from the 17th century (1670s). View on the Amstel, a river in the Netherlands, looking toward Amsterdam. The painting hangs in the Fitzwilliam Museum in Cambridge, England. *Photo is public domain.*

ABOVE: "Death wires" showing the orange balls barely visible on them. These are the wires that Elizabeth's husband, Alan Austin, nearly hit while flying a Jet Ranger helicopter through the Cahuenga pass that connects Hollywood and Universal City in southern California. *Source: Google Earth.*
BELOW: Patrick Williams conducting a symphony orchestra during a concert at Royce Hall, UCLA.

ABOVE: Albert Einstein playing violin on the S.S. *Deutschland*, a passenger liner, on his way to America in 1933. *Source: Mary Evans/Science Source, http://www.sciencesource.com.* BELOW: The Tacoma Narrows Bridge collapsing on November 7, 1940. *Source: University Libraries, University of Washington.*

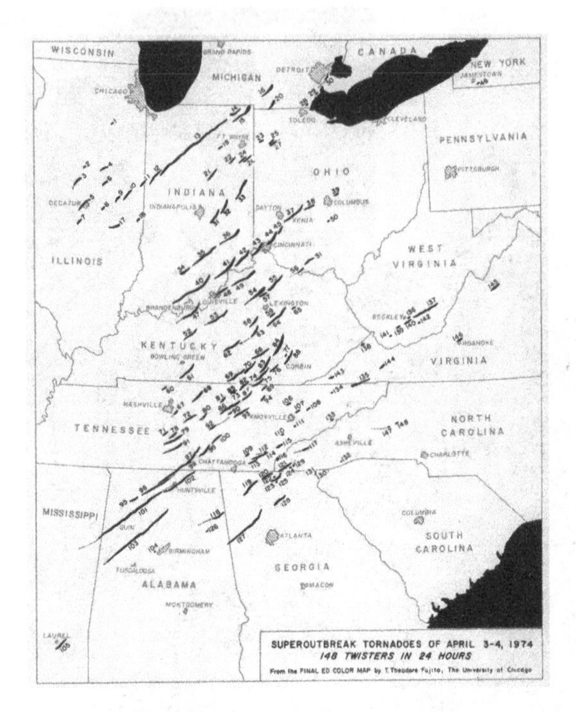

LEFT: Map of April 3–4, 1974, superoutbreak tornado tracks. There were 143 tornadoes in 13 states on these two days including 30 "violent tornadoes" rated Fujita Scale 5 (F5), the highest on the Fujita scale. There were 307 people killed and 5,454 injured. The damage from these tornadoes was estimated at U.S. $1.5 billion. If you added up the total path length of the tornadoes it was 2,521 miles. *Image Credit: Tetsuya Theodore "Ted" Fujita/University of Chicago.* BELOW: Wing tip vorticies created by an agricultural airplane is made visible using colored smoke rising from the ground. *Credit: NASA Langley Research Center.*

ABOVE: Lenticular clouds(altocumulus lenticularis), or mountain wave clouds. BELOW: A Foehn gap, or a cloud-free region downwind of a mountain crest between the cloud layers over the mountains and the large wall of clouds (on left) on the lee side of the Sierra Nevada mountains. The winds are blowing from the right to the left in this photograph, looking south down the Owens Valley in California. *Photo credit: Gordon Boettger.*

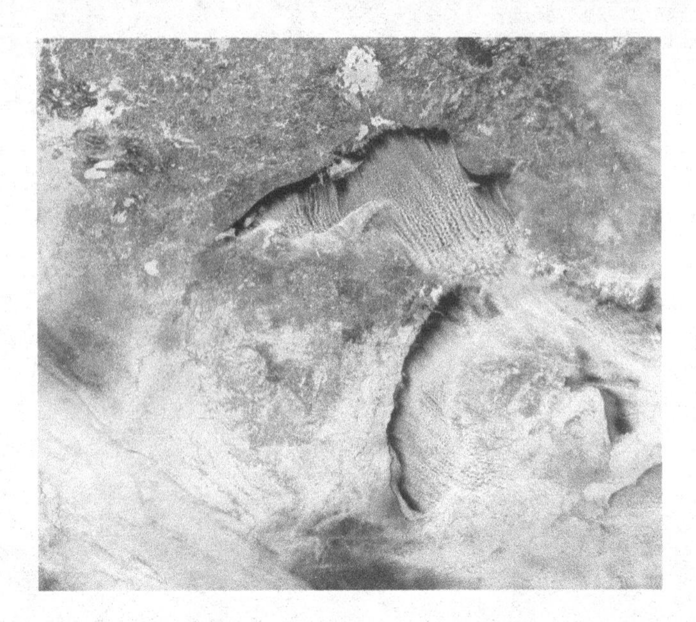

ABOVE: Cloud streets or long rows of cumulus clouds oriented parallel to the wind direction depicting turbulent convection in the atmosphere. The photograph was taken over the U.S. Great Lakes region by the NASA Moderate Resolution Imaging Spectroradiometer (MODIS) aboard a satellite. *Photo credit: NASA.* BELOW: A large dust storm that most likely originated in the Sahara Desert and swept across the Mediterranean Sea. The dust appears as a swath of tan over the dark open ocean and extends over Lebanon, Israel, the West Bank, Gaza, and Egypt. The photograph was taken on February 24, 2007 over the eastern Mediterranean Sea region by MODIS aboard NASA's Aqua satellite. *Photo credit: NASA. Annotated by WeatherExtreme Ltd.*

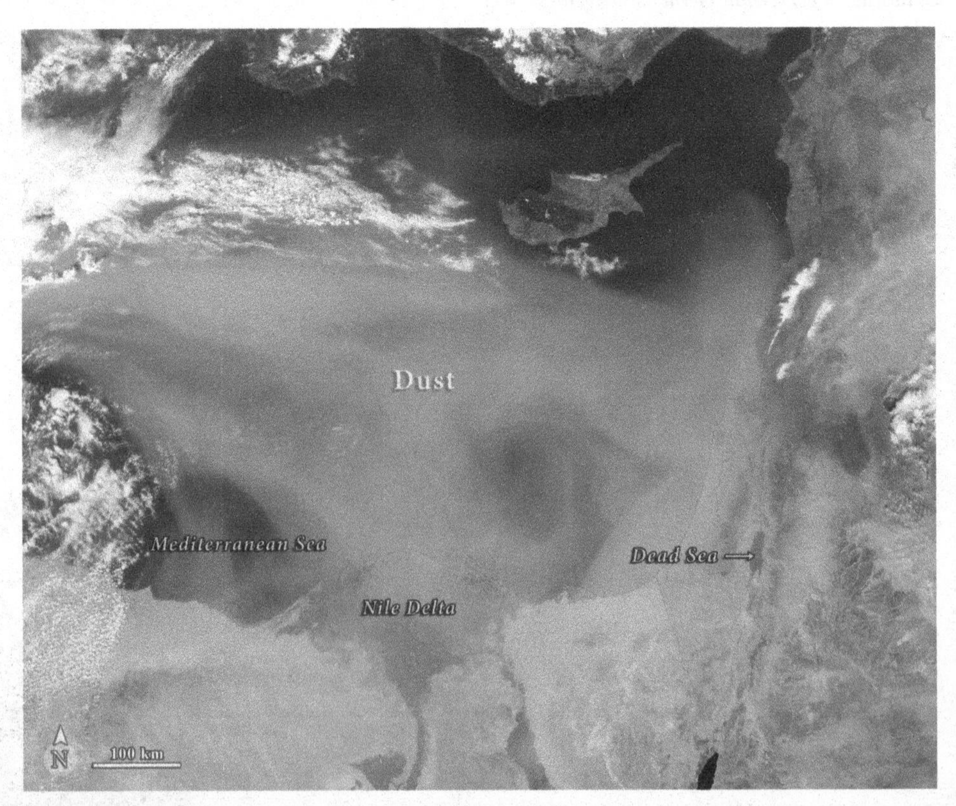

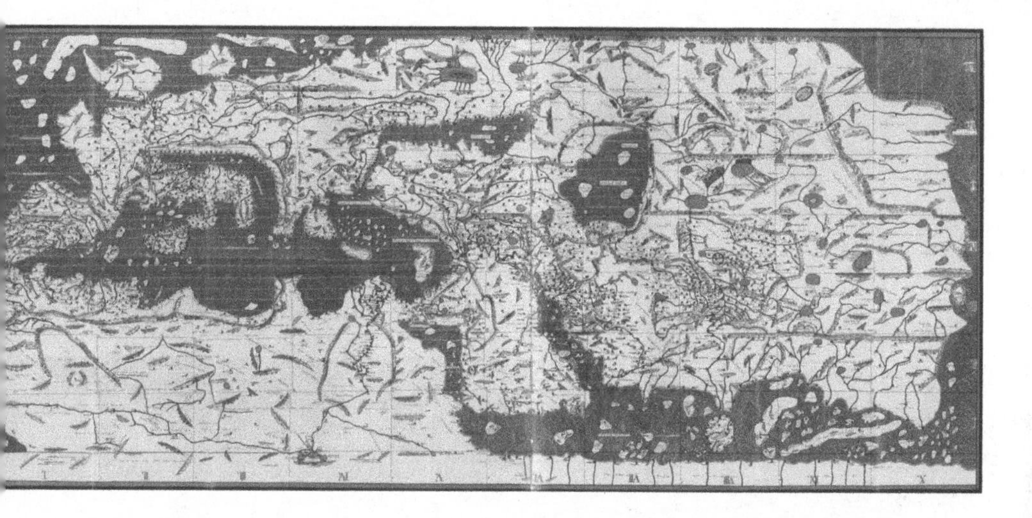

ABOVE: "Tabula Rogeriana" map of the world created by Muhammad "al-Sharif" al-Idrisi (c. 1100–1165), a Muslim scholar and geographer of the medieval Islamic period. His lineage was traced to the Prophet Mohammed. Al-Idrisi's maps were used for three centuries and inspired some of the world's greatest explorers. BELOW: Belay Demoz, Elizabeth Austin, and Edward Teets, Jr. (left to right). Photograph taken at an annual American Meteorological Society (AMS) Meeting.

Elizabeth backcountry snowcat skiing in the Sierra Nevada mountains of California.

ABOVE: Elizabeth Austin, Einar Enevoldson, and Steve Fossett (left to right) in Omarama, New Zealand. BELOW: Perlan I Team in Omarama, New Zealand. Left to right: Steve Fossett, Elizabeth Austin, Mike Todd, James Murray, Patricia Seamount, Michael Gallagher, William "Bill" Walker, Susana Conde, Einar Enevoldson, and Doug Hamilton.

ABOVE: Elizabeth standing beside Steve Fossett's Citation X jet while waiting to refuel in American Samoa. BELOW: Elizabeth and Robert "Hoot" Gibson, NASA astronaut, test pilot, naval officer, and aviator.

ABOVE: Elizabeth making notes from measurements made in a snow pit in the backcountry of the Sierra Nevada mountains to determine snow stability and avalanche probability. BELOW: Elizabeth and her mom and dad, Catherine and Patrick Williams.

ABOVE: Elizabeth standing in front of the United States Court House in Fort Worth, Texas, prior to testifying about weather conditions surrounding a plane crash. BELOW: Elizabeth, her husband, Alan and their son, Evan at the gondola stop at about 9,200 feet above sea level at the Heavenly Ski Resort, South Lake Tahoe. Ironically, this photograph was taken at almost the exact time that Steve Fossett was killed in a plane crash just south of Lake Tahoe, around 9:30 A.M. on September 4, 2007.

Elizabeth in one of the pilot's NASA pressure suits. The photograph was taken by Mike Todd, the Perlan I team and Red Bull Stratos life support expert.

ABOVE: Einar Enevoldson (left) and Steve Fossett (right) just after setting the world altitude record by a glider by soaring to 50,722 feet on August 30, 2006, in the Perlan I. The record was set out of El Calafate, Argentina. LEFT: Elizabeth in a helicopter on a forensic investigation site visit. They traveled to the location where a helicopter crashed into a reservoir.

ABOVE: Elizabeth preparing to give a weather forecast briefing to the Perlan I team in New Zealand.
BELOW: Elizabeth as a teenager on her beloved quarter horse, Vandy.

ABOVE: Harold “Hal” Klieforth, rear seat, and Joachim “Joach” Kuettner, front seat, in a Schweizer 2-25 sailplane. This is the same sailplane in which Joach set the altitude record of 42,700 feet above sea level on April 14, 1955, flying solo in the Sierra Nevada mountains of California. BELOW: Photograph from one of many of Kinsey Anderson’s atmospheric research projects at high latitudes. Inflation of a large polyethylene balloon at an airfield near Fairbanks, Alaska, in April of 1959. At sea level only about 1% of the balloon’s total expanded volume will contain helium. *Source:* Beneath Northern Skies: An Account of Research Carried Out at High Latitudes 1950 to 1959, *Kinsey A. Anderson, 2001.*

Kinsey Anderson (left) and his colleague, Donald Enemark calculating the amount of helium needed for a balloon inflation. This photograph was taken in Resolute Bay in the Arctic where they were about 125 miles east of the northern hemisphere's magnetic dip pole (where magnetic field lines are oriented vertically and plunge into the surface of the Earth). The theory at the time was that the very lowest energy cosmic rays might be able to reach the Earth's atmosphere at the dip poles (one in the Arctic and one in the Antarctic). This is one of the many atmospheric theories that Kinsey Anderson and his team were researching with their balloon launches. *Source:* Beneath Northern Skies: An Account of Research Carried Out at High Latitudes 1950 to 1959, *Kinsey A. Anderson, 2001.*

Crash of China Airlines Flight 642, McDonnell Douglas MD-11 in Hong Kong, China on August 22, 1999. The aircraft departed Bangkok, Thailand and was attempting to land at the Hong Kong International Airport during a typhoon.

ABOVE: Polar Stratospheric Clouds (PSC's), or nacreous clouds. Nacreous means pearlescene or pearl-like. These types of clouds inspired the name of the Perlan Project. (Perlan is an Icelandic word for pearl.) These clouds exist in the stratosphere and play an important role in ozone destruction. BELOW: *The Mothership.* A supercell thunderstorm that was stitched together from three photos to create this panoramic image. The photographer took the photos as the supercell traveled between the towns of Glasgow and Hinsdale, Montana. The supercell was between five to ten miles in diameter with winds around 85 miles per hour. *Photo credit: Sean R. Heavey*

ABOVE: A blow fly. Blow flies are an important tool in determining the time of death of a human body. BELOW: The Edmund Fitzgerald in 1971. *Source: www.inquisitr.com*

ABOVE: The Massachusetts Institute of Technology (MIT) Whirlwind computer developed to accommodate the demands of the SAGE system. This photograph was taken on the MIT campus in the early 1950s. *Source: MIT Lincoln Laboratory, http://www.ll.mit.edu/about/History/earlydigitalcomputing.html.* BELOW: A snow separator. The chimney assembly is the access point to the bins inside that collect the snow crystals. The hose on the left ducts the clean, filtered air in the plenum and chamber. *Photo courtesy Dr. David Mitchell, Desert Research Institute.*

ABOVE: The now decommissioned SAGE (Semi-Automatic Ground Environment) building located in Stead, Nevada, where the Desert Research Institute was located for many years. The building now houses the Nevada Terawatt Facility that is part of the University of Nevada, Reno. There are more windows in the building now, as seen in the photograph. These windows were installed by cutting into the three-foot thick walls in this massive 153,000-square-foot facility. *Photo courtesy University of Nevada, Reno.* BELOW: Diagram depicting the air temperature and moisture profile conducive to specific ice crystal type. Note that the "sweet spot" for snow crystal formation lies in the air temperature range between -10 and -20 degrees C (14 to -4 degrees F). *Source: Magono, C. and C. W. Lee. "Meteorological Classification of Natural Snow Crystals." Journal of the Faculty of Science, Hokkaido University, Ser. VII (Geophysics), II, No. 4 (November 1966), 321–325.*

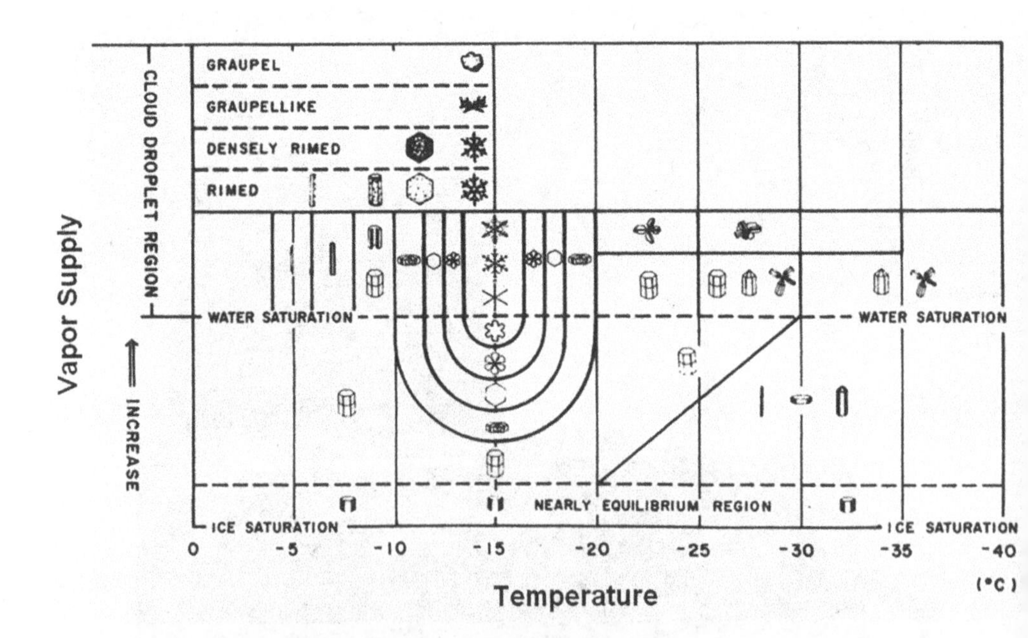

NOAA/NWS AND NASA'S SKY WATCHER CHART

High Clouds

Typical Types: Cirrus (Ci), Cirrostratus (Cs), Cirrocumulus (Cc)

H1: Cirrus
In the form of filaments, strands, or hooks

H2: Cirrus
Dense, in patches or sheaves, not increasing, or with tufts

H3: Cirrus
Often anvil shaped remains of a cumulonimbus

H4: Cirrus
In hooks or filaments, increasing, becoming denser

H5: Cirrostratus
Cirrus bands, increasing, veil below 45° elevation

H6: Cirrostratus
Cirrus bands, increasing, veil above 45° elevation

H7: Cirrostratus
Translucent, completely covering the sky

H8: Cirrostratus
Not increasing, not covering the whole sky

H9: Cirrocumulus
Alone or with some cirrus or cirrostratus

Middle Clouds

Typical Types: Altostratus (As), Altocumulus (Ac), Nimbostratus (Ns)

M1: Altostratus
Mostly semi-transparent, sun or moon may be dimly visible

M2: Altostratus or Nimbostratus
Dense enough to hide the sun or moon

M3: Altocumulus
Semi-transparent, one level, cloud elements change slowly

M4: Altocumulus
Lens-shaped, or continually changing shape and size

M5: Altocumulus
One or more bands or layers, expanding, thickening

M6: Altocumulus
From the spreading of cumulus or cumulonimbus

M7: Altocumulus
One or more opaque layers, w/ altostratus or nimbostratus

M8: Altocumulus
With cumulus-like tufts or turrets

M9: Altocumulus
Chaotic sky, usually at several layers, maybe w/ dense cirrus

Low Clouds

Typical Types: Stratus (St), Stratocumulus (Sc), Cumulus (Cu), Cumulonimbus (Cb)

L1: Cumulus
With little vertical extent

L2: Cumulus
Moderate/strong vertical extent, or towering cumulus

L3: Cumulonimbus
Tops not fibrous, outline not completely sharp, no anvil

L4: Stratocumulus
From the spreading and flattening of cumulus

L5: Stratocumulus
Not from the spreading and flattening of cumulus

L6: Stratus
In a continuous layer and/or ragged shreds

L7: Stratus Fractus and/or Cumulus Fractus
Of bad weather

L8: Cumulus & Stratocumulus
Not spreading, bases at different levels

L9: Cumulonimbus
With fibrous top, often with an anvil

NOAA
NATIONAL OCEANIC AND ATMOSPHERIC ADMINISTRATION
U.S. DEPARTMENT OF COMMERCE

NASA

ABOVE: Cloud classification chart depicting some of the main types of clouds. *Source: NOAA/NASA.*
BELOW: Sample of a fractal image. This image shows a portion of what is known as the Mandelbrot Set. *Source: Peter Alfeld, Department of Mathematics, University of Utah.*

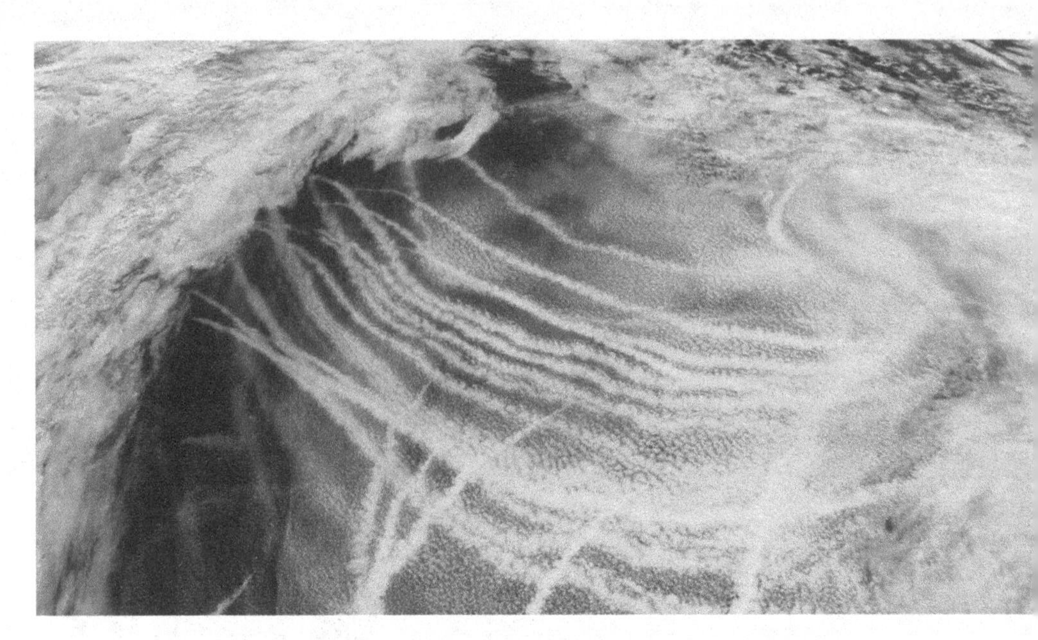

ABOVE: Image of ship tracks out over the ocean. Ship tracks form in the wake of ships and are composed of cloud water that condenses on the fine particulate matter in the ship's exhaust when the air is supersaturated (i.e., with relative humidity greater than 100%). *Source: NASA Moderate Resolution Imaging Spectroradiometer (MIDOS) flown on NASA spacecraft.* BELOW: Mammatus clouds over Austin, Texas. These rare clouds show the locations of sinking air (the pouch-like structures) and are generally seen after the strongest portion of a thunderstorm has passed through the area. Mammatus are usually seen on the underside of a thunderstorm's anvil although they can be found elsewhere too. *Photo credit: Matt Roberts, Public Domain. http://public-domain.pictures/view/image/id/18119496525#!Austin+Mammatus+Cloudpocolypse.*

THIRTEEN

La Calamidad de México

If you don't change your direction, you may end up where you are heading.

—Lao Tzu

We had finally decided to make a big move . . . to Mexico. Cabo San Lucas would be our new home. The opportunity to live in a wonderful climate and run our business, still staying relatively close to the United States, sounded great to us—an adventure. Alan and I had never taken a vacation together or even had a honeymoon, for that matter. This was a chance to combine some recreation, adventure, and work, all together. If we were going to be workaholics, at least we wanted to do it in a sunny, warm beach community.

Cabo San Lucas, or "Cabo," is a city of approximately 68,000 and lies on the southern tip of the Baja Peninsula. On one side of the peninsula is the

Pacific Ocean and on the eastern side lies the Gulf of California. Cabo lies just south of the Tropic of Cancer and is almost identical in latitude to Havana, Cuba. It is mainly a tourist destination where beaches, surfing, fishing, and spas are the main attractions. The climate in Cabo is that of tropical desert. In the winter the prevailing winds are westerly, bringing with them cool and dry Pacific air. During the summer the trade winds shift and bring in higher humidity and storms from the east and south. The summer temperatures can get quite hot, but it is still relatively dry; compared to much of the United States, the heat is bearable. It does not feel nearly as hot, for example, as summers on the East Coast. The ocean temperatures at Cabo range from 70 to 72 degrees Fahrenheit in the winter and 82 to 84 degrees F in the summer, and can at times rise above 90 degrees F.

Alan and I had planned it all out as thoroughly as we could possibly imagine. We had done a tremendous amount of research, hoping to avert any unpleasant surprises from this drastic change in scenery as we traveled the 1,300 miles from Lake Tahoe to Cabo with our young son. But our plan was to keep our Lake Tahoe house as a base in the United States so that we could go back and forth to the States seamlessly as needed. At least this was the plan. Our company, WeatherExtreme Ltd., would remain the same U.S. firm, as I traveled so much anyway, it really didn't matter where we were at any particular time, while everyone else in the company remained in Stateline (Lake Tahoe), Nevada.

"El jefe no está aquí hoy y es el único que puede estampar y puede firmar su pasaporte." (The chief is not here and he is the only one who can stamp/sign your passport.) The man told me this as I stood there with my *Forma Migratoria 3* (FM3) in hand. An FM3 is a type of visa that is required by the Mexican government if one is to live in Mexico. There are other types, including an FM2, which Alan had, which was a working visa. I was in the Instituto Nacional de Migracion in downtown Cabo, where one must get his or her visa stamped in order to leave Mexico, if you're there on an FM, or a residential visa. *Why didn't I know about this before?* I thought to myself as I became more exasperated.

"Pero debo estar en los Estados Unidos por este el jueves. (But I *need* to be in the U.S. by this Thursday)," I replied, butchering the beautiful Spanish

language. The man explained that not only did I need to have *El Jefe* sign and stamp my visa but also that at least one week advance notice was required! On top of it, it turns out that *El Jefe* comes into the office only when *El Jefe feels* like it. Today I think back on our Mexico experience and am reminded of a Robert Burns poem, "To a Mouse." The particular verse I think of is:

But, Mousie, thou art no thy lane,
In proving foresight may be vain;
The best-laid schemes o' mice an' men
Gang aft agley,
An'lea' us nought but grief an' pain,
For promis'd joy!

Or as it's commonly referred to: "The best laid schemes of mice and men!"

I finally made it to the United States after I begged and pleaded to get someone other than *El Jefe* to sign my FM3. I arrived in the Midwest where I was to give a deposition regarding the tragic death of young girl. Was it a tornado or a gust from a severe thunderstorm that caused her death? To complicate things further, straight-line winds can be an element of tornadoes at times. Turns out, for insurance and legal purposes it didn't matter. But this case brought to light a few problems we had had in the past about determining causes of events. In the past, weather forecasters and meteorologists from our National Weather Service would go out and survey the damage. They alone determined damages and causes to structures and whether it was a tornado or not. No engineers, no contractors, only meteorologists. Luckily, this practice has since changed. Analogously, it would be like having a personal trainer from the gym determine the cause of death while on a Pilates machine of one of the club's regulars. The outcome of the tornado case, though it was settled, remains unclear as to whether it was a tornado or straight-line winds.

The bottom line is that these extreme events kill many people and the necessity for timely and accurate advance warnings is critical. But these warnings must be used in conjunction with a plan. Chris Miller, the

Warning Coordination Meteorologist at the National Weather Service in Lincoln, Illinois, stated it perfectly, "If a business or individual or school or hospital doesn't have a plan, we could put out the best warnings in the world and it's not going to help anybody." It is the combination of warnings and planning that saves lives. Recently there have been huge advances in the timeliness and accuracy of tornado warnings. Much of this success is owed to Doppler weather radar technology and its ever-increasing advancements.

"Tornado" is a Spanish word, *tronada*, which means thunderstorm. When we hear of tornadoes occurring, we most always think of the U.S. Midwest or "tornado alley," which stretches from central Texas to Nebraska in the Central Plains region of the U.S. Why does tornado alley exist and do tornadoes occur anywhere else in the world? The U.S. experiences around 75 percent of all tornadoes that occur globally. The majority of these occur in the summertime in the Midwest, although every state in the U.S. has experienced a tornado, even Alaska and Hawaii, at some point. The next most common region for tornadoes is from central Texas east over through Alabama. Canada comes in second behind the U.S. for tornado frequency, with 5 percent of all global tornadoes occurring there. Mexico does not experience very many tornadoes at all since it does not get the cold air intrusions into its mostly tropical climate to produce them. Only very northern parts of Mexico have reported tornadoes in the past. Behind Canada is Russia, and then the United Kingdom. Then there is Europe's tornado alley—although very few tornadoes occur there—which runs from northeastern France through Germany and into Poland. Other locations around that world that report tornadoes are South Africa, parts of Australia, New Zealand, eastern Argentina, southern Brazil, and portions of Asia.

The recipe needed to form a tornado is that there must be contrasting air masses, moist and hot combining with cold and dry. Then strong vertical wind shear must exist, so the air at the surface is forced upward, forming thunderstorms. Then a change in wind direction and an increase in wind speed with altitude creates a spinning effect in the thunderstorm. This area of rotation is what can spawn a tornado. Tornadoes can be generated from non-thunderstorms (i.e., convective storms, without lightning and thunder).

Tornadoes can develop in any number of severe weather situations. But the most dangerous tend to form from supercells. Supercells are severe thunderstorms that contain a deep and persistent rotating updraft called a mesocyclone (NOAA definition). The mesocyclone is a vortex of air within the supercell that rises and rotates around the vertical axis, thus the propensity for producing strong and destructive tornadoes in addition to damaging winds and large hail. Mesocyclones are usually two to six miles in diameter.

The Central Plains of the Midwestern United States has certain conditions that make it ripe for the development of severe thunderstorms that can then spawn tornadoes. The warm, moist air that flows from the Gulf of Mexico combines with the colder, drier air flowing down from Canada. When cold, dry air aloft moves over warm, humid air at the surface, this then produces an atmosphere that is termed "conditionally unstable," which means that the atmosphere is unstable for a moist parcel of air being lifted but stable for a dry air parcel. Then the presence of a strong upper level jet stream adds the strong vertical wind shear required to get the rotation going. And a tornado is formed.

In the U.S. and Canada, tornado intensity is measured on a scale called the Enhanced Fujita Scale (EF Scale). But this scale is based on wind estimates and not measurements based on damage. The scale goes from EF0 through EF5, where an EF0 tornado has three-second wind gusts estimated to be 65 to 85 miles per hour, which would cause minimal to no damage. On the other end of the scale, an EF5, the strongest on the scale, the three-second wind gust range is over 200 miles per hour and can cause total destruction. Dr. Tetsuya Theodore "Ted" Fujita, from the University of Chicago, first introduced the original Fujita Scale back in 1971 and it has now been used to categorize every tornado in the U.S. since 1950. Dr. Fujita also discovered microbursts and macrobursts.

Most of the tornado destruction in the U.S. and Canada occurs in the summer months and most often between the times of three P.M. and nine P.M., when heating of the day is at a maximum. A typical tornado lasts only for a few minutes (five to ten minutes on average), and rotates counterclockwise in the northern hemisphere and clockwise in the southern hemisphere due to the coriolis force, an artifact of the Earth's rotation. Tornadoes vary greatly in their size and intensity, but the average path length is nine miles long and two hundred yards wide. The U.S. averages eight hundred tornadoes per year of average size, but there can be "super outbreaks" of tornadoes, which can be extremely deadly. There was a super outbreak on April 26–28, 2011, when there were a whopping 293 tornadoes in twenty-one states. Three hundred twenty-four people were killed, but this super outbreak is ranked second, due to the fact that there were fewer "violent" (EF4 or EF5) tornadoes, in terms of damage and destruction, with fifteen. The worst super outbreak in American history occurred on April 3–4, 1974, when there were thirty extremely violent tornadoes, with 307 casualties. Both of these super outbreaks had tornado path lengths totaling over 2,500 miles. The tornado outbreak of 1953 caused the highest number of fatalities, when 519 people were killed by three individual tornadoes (May 11 in Waco, Texas; June 8 in Flint, Michigan; and June 9 in Worcester, Massachusetts).

The advent of weather radar and particularly Doppler radar has helped meteorologists in forecasting tornadoes and potential tornado outbreaks tremendously, which in turn helps with preparedness and evacuations, although as the high death toll of 2011 shows, there is still a lot to be done.

Having a plan and being prepared for tornadoes is critical, as time is of the essence when a tornado strikes. Going underground is the safest option, but if this option is not available, having a safe room in a house or business is the next best choice. If you find yourself outside as a tornado is approaching, take cover as best you can, and if there is a vehicle nearby, get into it, buckle up, and keep your head down. It is a terrible idea to try to drive as fast as one can to try to get away from a tornado. Although cars travel faster than tornadoes, one can hit traffic jams, must follow roads, obey the laws, and try not to get into an accident all while trying to outdrive a tornado. In addition, flying debris can cause serious injuries if not death.

The term RADAR stands for Radio Detection And Ranging. By the end of World War II, radar had been highly developed and successfully used. The purpose of radar at this time was primarily for locating aircraft during the war. But then using radar for use in weather forecasting took hold. There are many types of radar, from pulsed radar, wind profilers, Doppler radar, aircraft surveillance radar, to continuous-wave radar.

Radar comes in many configurations and types, most of which the general public more than likely understands. There are detection and search type radars. These types of radar scan and search large areas with pulses of short radio waves. Ships and planes are generally (not taking into account here the new burgeoning stealth composite technology) made of metal and reflect these radio waves. The radar measures the time from emission to reception to calculate distance.

Then there are targeting radars, used for such things as missile guidance. This type of radar is similar to detection radar, but is designed to cover a much smaller area, and used much more often with the goal of acquiring a target and locking on it quickly.

There are navigational radars, like the type used by air traffic control. Air traffic control and navigation radar uses primary and secondary radars. Primary radars are the type of radar that reflects all kinds of echoes, including aircraft and clouds. Secondary radar emits pulses and listens for an answer of digital data emitted by an aircraft transponder. Transponders send out different varieties of data. For example, the military uses transponders to establish the nationality and intention of an aircraft—always looking for hostile radar returns.

I think many people were amazed at the complete disappearance of a 777 Boeing aircraft, Malaysian Airlines Flight 370. How can that happen with today's radar technology? In reality, most of the world is *not* covered by radar, and therefore aircraft tracking has to take a different form. Aircraft that are not in radar contact have to make what are called position reports. These voice reports must be in a prescribed form that indicates altitude, position, speed, and the like . . . and also give an estimate of the next reporting point. These communications are possible through the use of remote center air/ground sites composed of both VHF and UHF transmitters and receivers. Pilots are directed to change to various discrete frequencies to make subsequent reports as they move along their routes around the globe. These reports are for the purpose of identifying and tracking aircraft, not necessarily for weather. Most aircraft that fly in non-radar environments around the world on a daily basis do have weather radar on board, and GPS is rapidly changing this lack of tracking radar for the better, in terms of more accurate and up-to-date aircraft tracking and separation in these radar-less zones. Alan told me that during flights in the Caribbean and South America that he would often not even get a response from air traffic controllers. After making a position report while he was flying from Quito, Ecuador, to Rio de Janeiro after refueling in Manaus (approximately 1,700 miles from Rio), he continually made position reports and attempted to reach a controlling authority on the radio, but he didn't receive any response until he was on short final for an Instrument Landing System (ILS) approach to the airport. He said, "I'm two miles out for ILS twenty left," when the tower calmly replied, "Cleared to land." It was the first response he had gotten in almost 1,700 miles. Flying around the world outside the United States can sometimes be quite a challenge.

There are literally dozens of other radars, of course, but this isn't intended to be an exhaustive treatise on radars, rather just a simple explanation and clarification of the basics of the most common and most pertinent to the weather. The most pertinent type of radar is weather-sensing radar. Weather radars, to some extent, resemble search radars. They also use radio waves, along with various types of polarization. The frequencies of weather radar are employed to cut through precipitation, reflectivity, and attenuation (when the radar signal weakens due to energy loss from scattering and absorption)—a function of atmospheric water vapor.

When Doppler technology was added to the weather radar, forecasters were able to obtain even more information about storm systems and the state of the atmosphere. Christian Doppler, an Austrian physicist, proposed the effect that bears his name back in 1842. Most everyone is familiar with this effect, as it is an increase (or decrease) in the frequency of sound, light, or other waves as the source and the observer move toward (or away) from each other. If a police car is driving toward you and then passes by, the siren sounds higher as it approaches, and then as the car moves away from you, there is a noticeable change in the pitch of the siren. It sounds lower as the car moves away from you. This is an example of the Doppler effect. For Doppler weather radars, this effect is used to measure the velocity of particles, such as precipitation particles, in the atmosphere. We can also measure the volume of particles in a particular location and measure their radial velocity as well. Radial velocity is the velocity of the particles toward or away from the radar antenna. These radars also scan in a 360-degree rotation around the radar antenna and can also make scans above the horizon at various elevation angles, thus allowing forecasters to see storms and weather events. The radar sends out short electromagnetic pulses of energy and measures the power received from the target as the pulses bounce back. It also measures the speed of the target either toward or away from the radar source. Today the National Weather Service has 160 NEXRAD Doppler radars in place and most of them have been recently converted over to dual-polarization, meaning they can now identify different precipitation types, for example. These NWS radars have a wavelength of ten centimeters, or S-Band part of the microwave band of the electromagnetic spectrum, radars. There are also around forty-five Terminal Doppler Weather Radars (TDWRs) located near airports. The primary purpose of these C-Band (or five-centimeter) radars is the detection of hazardous weather at and around airports. The TDWRs give rapid updates and provide higher resolution readings to coincide with the readings from the NEXRAD radars. There are many more types of weather radars used in atmospheric research including K-Band, X-Band, and W-Bands, depending on what the users are trying to detect and analyze.

In forecasting for the possibility and probability of tornadoes, meteorologists rely heavily on NEXRAD Doppler radars to look for and identify features that might indicate tornadic activity, such as hook echoes, tornado vortex signatures, or the presence of a building mesocyclone. These radars even have algorithms that can calculate hail probability, storm structure, and vertically integrated liquid, all of which can occur in tornadic activity.

Back in 2008, Dr.'s Kevin Simmons and Daniel Sutter, both economists, published results of their investigation of the correlation between tornado warning lead times and the number of casualties. Though their results were mixed, their most important finding was that an increase in lead time in warnings—up to fifteen minutes—reduced fatalities, whereas lead times longer than fifteen minutes actually increased fatalities, versus when there were no warnings at all. The "false alarm" rate during their study period for lead times of up to fifteen minutes was 75 percent, and was 95 percent if you looked at the occurrence of a tornado hitting within the fifteen-minute period after the warning was issued.

If people think it is a false alarm, they do nothing, and thus the result is the same number of fatalities as if there was no warning at all. So there is obviously quite a bit of improvement that can be made, and these new dual-polarized NEXRAD Doppler weather radars are just what the doctor ordered: they provide much more detail of a multitude of parameters sooner and at a much higher resolution that data meteorologists utilize in their analysis of possible severe weather situations, allowing them to provide more accurate forecasts much more quickly.

—

When I finished my deposition testimony regarding the little girl's death and headed to the airplane to catch my flight back to the Los Cabos International Airport, I found that all of the flights to Cabo were canceled and the airport was closed due to weather. The irony was quite rich. A tropical storm was traveling directly over Cabo and directly on up the Baja Peninsula. I wouldn't be able to get home for days. So I quickly changed plans and flew to southern California and stayed with my parents to wait out the storm. Alan and Evan remained in Cabo and rode out the storm, which resulted in two fatalities. After three days of waiting, the Los Cabos Airport reopened and I was able to get home.

Even being a meteorologist, I did not quite realize the number of tropical storms and hurricanes that hit the Pacific. The Pacific Ocean region is actually the second most active hurricane region in the world, behind the Atlantic. On average, nine hurricanes per year form in the Pacific, with four of these being major hurricanes. Dr. Jim Means, one of WeatherExtreme Ltd.'s atmospheric scientists, is studying hurricanes that can make it to San Diego and the chances of future hurricanes hitting this region as climate change continues to alter weather patterns. Though tropical storms and hurricanes in the Atlantic that affect the Caribbean and eastern United States are still the most deadly and newsworthy, these Pacific storms are growing in frequency and strength. Changing weather patterns will come to affect not just hurricanes in the Pacific and the Atlantic, and the effect they will have on these highly populated regions, but they may also affect the dispersal of tornados in the heartland, although the link between climate change and tornadoes is still unclear.

Back in Mexico, the weather had cleared up—it was beautiful. Unfortunately, Alan and I came to realize that there was a tremendous difference between retiring or vacationing in Mexico and doing business there. It had become painfully obvious that there were too many insurmountable issues making impossible for us to continue our life there. It wasn't meant to be, so we chalked it up as an interesting experiment and returned to the States.

FOURTEEN

The China Factor

For a successful technology, reality must take precedence over public relations, for Nature cannot be fooled.

—Richard P. Feynman

I was listening to a presentation in Chinese through a translator via a pair of headphones. The presenter was speaking in Chinese, and a translator in a booth was simultaneously translating what was being said into English. During the presentation, the translation into English would continually have unexplainable gaps and pauses, and it did not seem like they were due to the translator's proficiency. The gaps and pauses continued. *What is going on?* I thought to myself. I was one of ten experts from various fields invited to Harbin, the capital of the People's Republic of China's most northeasterly province, Heilongjiang, by the United Nations Development Program, to each give a presentation in our area of specialty in order to help the Chinese in their pursuit of building 350 new ski resorts as close to Beijing as nature would allow, given the warmer temperatures and terrain issues closer to Beijing. I'm surprised I didn't realize what was happening sooner. The government operators, who were staffing the AV feed, did not approve of what the presenter was saying and cut off the audio translation to the headsets of the four hundred or so attendees in the auditorium as occasion warranted. Censorship was alive and well in China.

The list of attendees was impressive and included most of leaders of the Heilongjiang People's Government, the deputy director of China's national tourism agency, the mayor of Harbin, the chairman of China's Skiing Association, most of the heads of tourism for the different provinces of China, the Italian ambassador to China, many leaders of the United Nations Industrial Development Organization, various foreign and Chinese newspapers, and many Chinese television and radio representatives.

The Heilongjiang Province lies in the northeastern part of the country and has a population of over 38 million. It lies 630 miles northeast of Beijing and 1,750 miles northeast of Hong Kong. "Heilongjiang" means "Black Dragon River" in Chinese. This river is also known as the Amur River and is the tenth longest river in the world, at 1,755 miles long. The Amur headwaters begin in western Manchuria and the river flows east and southeast with much of the length of the river defining the border between Russia and China. The river ends in Russia, where its mouth is at the Strait of Tartary, the strait that connects the Sea of Japan and the Okhotsk Sea. There are many fish species endemic to the river, including sturgeon, carp, and the largest fish found in the Amur, the kaluga, which can reach eighteen feet in length. The Amur River has been continuously contaminated over many years by both China and Russia. The Chinese side has contaminated the river through chemical plant explosions and other massive toxic releases into the river. The Russian side has contaminated it through, among other things, poor mining practices. Both sides contaminate the river through wastewater discharge, especially on the heavily populated Chinese side. This does not even take into account some of the farming practices that pollute the river with agricultural runoff. Some of the pollutants include benzene, mercury, anthracene, and pyrene. The people in the region suffer health problems, including cardiopulmonary disease and respiratory problems, due to consuming the polluted water and eating the fish caught from the river. The ecosystems suffer, too, as there are many species of birds, plants, wetlands, and trees along the Amur whose future is severely endangered.

But I was not in the city of Harbin to visit the Amur River. In fact, the city of Harbin lies along a different river, the Songhua River, the largest tributary of the Amur River. I was here as a consultant to the Chinese. The United Nations had invited ten experts—and I was the only American, and the only woman in the crew—to give presentations and meet with the Chinese during their International Skiing Industry Cooperation Forum. To those few who may be unaware of it, skiing is big business. It also requires careful planning and concern for the environment—or perhaps better put, *should* include concern for the environment. In China most of the people live in the south, where the climate is warm, and most of the potential skiing is in the north. It gets so cold in this region of China that Harbin is also known as the Ice City—they even have a very famous ice festival every year, with some amazing ice sculptures and other related festivities. Winter tourism, of course, also includes a ski resort.

The purpose of our group's visit to this region of China in December was for each of us to present our area of expertise, with the ultimate goal being to provide assistance in choosing the locations of the ski resorts, determining what types of infrastructure were required, and what locations just did not warrant the construction of a resort. During the month of December the average high temperature for the city of Harbin is 16 degrees F and the average low temperature is –4 degrees F! But it was even colder during my visit. It certainly lived up to its reputation of producing an abundance of ice. Prior to my arrival in China, I had heard that the Chinese wished to build as many as three hundred fifty new ski resorts, and the closer they could be built to Beijing, where more of the population resides, the better. But the farther south you go, the warmer it gets, and thus there is tremendous need for snowmaking at any ski resorts to be built in these more populous climes. Snowmaking is a complicated and expensive process that begs to be considered carefully when looking at potential ski resort locations. That is where my expertise came in. I was there to talk about the weather and climate in regard to ski resort development and snowmaking operations.

I presented data and information about effective snowmaking and the importance of ski run location in addition to the ski resort location. For example, if a ski run is cut out of the forest such that the prevailing winds flow up the ski run, this makes it difficult to make and keep snow on that run. In this example, if changing the orientation of the ski run to the wind is not an option, then leaving an "island" of trees at various locations periodically down the center of the run can help keep and maintain the snowpack on the run, by providing wind block and also acting as snow fences that reduce the wind speed and cause snow to drift and be deposited behind (downwind) of these tree islands.

I also discussed the necessity and critical importance of installing weather observing stations at various elevations and slopes on the mountain. Monitoring the weather conditions is of utmost importance for good snowmaking. There are critical temperatures for snowmaking and thus forecasting for and monitoring of these is essential, as the wet bulb temperatures (which take the air temperature and humidity into account) are what must be measured and monitored. Snowmaking also depends largely on the type of snowmaking systems implemented, as some use a combination of air and water that is spewed out of a snow gun, while others use large fans instead of guns to combine the water and air to create snow particles.

What is most interesting, but perhaps not surprising, is that in China the government plays a large role in the development of the skiing industry. The land and forests where the resorts are built and new ones are planned to be built are mostly government owned, and the capital funding the construction is mostly government controlled, too.

After giving my presentation and some interviews to the press, I walked back upstairs in the Sinoway Hotel in Harbin, where the forum was taking place, to change clothes. I had left one of my blouses on the bed because it needed dry cleaning, but I could not find those bags that are usually in the closet for dry cleaning to be put out for the maid service. Not five seconds after I walked into my room the phone rang. It was the maid service asking if I wanted my blouse cleaned! How did they know? They must have seen it earlier when they cleaned the room, I thought to myself. I told her that yes, I wanted my blouse dry cleaned. After hanging up the phone, I wondered how they even knew I was in my room, as I hadn't passed anyone in the hallway. It was a strange and uncomfortable feeling.

We were to be picked up at the hotel the next morning to attend the grand opening festival at the Longzhu Erlongshan Ski Resort, also known as the Double Dragon Mountain Ski Resort, located outside of Harbin. Now there are five resorts in this Yabuli region of China, but the Double Dragon is still China's largest ski resort and considered it best. In the morning, I left my hotel room, where the water always had an unpleasant odor. I would never figure out exactly why, but my thought is that it is due to the extreme pollution in the Amur River contaminating the potable water supply. I headed downstairs to meet the other nine in our UN group for our trip to the ski resort. A couple of the people brought their ski gear. I was suddenly jealous that I had not brought mine. We were loaded into a small bus and started our trip to the resort, which was located almost forty miles east of Harbin. We had an entourage of security people and escort vehicles that shadowed us everywhere, and this trip was no exception. For whom were the security people, I asked myself frequently, while thinking of the censors that kept clicking off my headphones at the first presentation we attended. Harbin is big, a city of ten million people, and yet anytime we traveled anywhere we had official police and security vehicles in front of us, behind us, on the sides, and at every intersection stopping traffic for us. We never, ever, stopped at one traffic light! Oh, the stares we got from the locals! I felt like Princess Diana being paraded around and shadowed everywhere she went across the globe, and not in a good way.

As we drove, we saw men and women at various locations sweeping the sides of the road with brooms. But these weren't brooms like we are used to seeing, but rather large sticks with small sticks tied to the bottom of the larger stick. They didn't look very effective and it seemed as though it was a way to provide jobs for people. I had the same thought when once in Paris I saw the men in the bright green jumpsuits who sweep the streets using that exact same ancient technology.

We finally arrived at Longzhu. The resort is not large, with just eight trails, but it covers 192 acres with its hotels, restaurants, shops, and areas for activities such as snowmobiling and sledding. We were approximately 275 miles north of the border with North Korea and it was freezing out.

We spent more than an hour standing outside in the cold but clear blue sky weather, watching wonderful shows put on by children and adults dressed up in the most colorful outfits and costumes. There were shows with dragons that were made up of five or six adults underneath the costume, and people playing musical instruments. It was colorful and fabulous. But it felt great to go inside the ski lodge when it was over and have a hot, albeit strange, meal. There were people out skiing and some ski instructors taking runs to show off the resort for us guests. I had taught skiing while putting myself through graduate school and could tell by the way they skied that these ski instructors were trained by Europeans and not Americans, who have a bit of a different style of skiing, which is instantly recognizable to another ski instructor but difficult to put into words, as it is the manner in which they move their bodies. The whole trip was a wonderful but very strange experience, and not likely to be the last of such trips for weather and environmental experts to the country.

The economic opportunities in China are growing exponentially but so are their environmental troubles. With the largest population in the world of almost 1.4 billion, followed by India at almost 1.3 billion, and then the United States at 324 million, the stresses they create on the Earth (population over 7 billion) are also growing exponentially. At a minimum, these stresses include maintaining safe drinking water, clean air for breathing, and healthy soil for a sustainable level of crops and livestock to feed everyone. In addition to these more immediate threats due to continued population growth, there is a potentially bigger one that could exacerbate these issues still further: global climate change. Yes, the climate of our planet is changing. It has in the past and it will continue to change in the future. The issue today is that humans are causing some of this change. Though scientists debate about how much humans contribute to this change, it must be acknowledged we are very likely responsible for the most recent increase in global average surface temperature. There is no doubt, however, the climate is changing and that this change threatens our world as we know it.

Because there is obviously a lot of misunderstanding going on in this arena, I think a little background is in order here. Climate and weather are different. Weather is the condition of the atmosphere at a particular time and place, whereas climate is an average of weather characteristics. A little saying to remember this is: "*Climate* is what you *expect*, but *weather* is what you *get*." Both weather and climate issues must be addressed, but it is much easier for the general public to take action against weather events as opposed to climate ones, as weather events exist here and now and are usually destructive in some manner. Climate issues, on the other hand, can easily be put on the back burner, so to speak. We all have busy daily lives with work, family, health, and finances: climate change does not enter many people's thoughts on a daily basis.

Our "climate system" not only involves the atmosphere but also the land, ecosystems, snow, ice, oceans, rivers, lakes, and streams. Scientifically speaking, the climate system is comprised of: the atmosphere, the cryosphere (water in solid form), the hydrosphere (water), the land surface, and the biosphere (all ecosystems including us humans, plants, other animals, and microbes). This system is driven by the energy from the sun, some of which is reflected and some of which is absorbed by the Earth and the atmosphere. The sun's energy obviously varies greatly around the globe and is not uniform over Earth, thus we experience a wide range of weather across the planet. But when the balance of energy on Earth (incoming and outgoing) is disturbed, this changes the climate system.

I prefer the term "climate change" as opposed to "global warming," as I feel the latter term can be a bit misleading if you take our atmosphere as a whole (and not just the troposphere). Yes, global average surface temperatures are warming, but our stratosphere is actually cooling, and this cooling can cause further destruction of the ozone by creating a larger ozone hole. Also, it can be misleading to people who confuse global averages with the temperature at their particular locale. Though some places may be experiencing colder average temperatures, our globe as a whole is warming at the surface and in the troposphere. Climate change also extends to the oceans, which are not immune to these effects, either. Overall, their salinity is increasing (i.e., becoming more acidic), ocean temperatures are warming, and sea levels are rising. Why? That is the million-dollar question.

First, what is our atmosphere composed of, anyway? At the surface it is 78 percent nitrogen, 21 percent oxygen, 0.9 percent argon, and 0.1

percent trace gases. It also contains between 1 and 4 percent water vapor, depending on where you are on the globe and what time of day it is. There are also traces of dust particles, pollen, and other solid particles. Though they only comprise 0.1 percent of the total, the trace gases are very important players in climate change. Breaking this 0.1 percent trace gases bracket down, we have, rounded off: 93.5 percent carbon dioxide, 4.7 percent neon, 1.3 percent helium, 0.41 percent methane, 0.08 percent nitrous oxide and 0.01 percent ozone. Though these gases are in trace quantities in our atmosphere, some of them have a mighty presence in terms of our climate.

The current science shows that the troposphere is warming slightly more—and much more uniformly—than the surface climate. Meanwhile, the stratosphere is actually cooling. Yes, climate has fluctuated tremendously over the lifetime of the Earth, but there is no doubt that warming is presently occurring where we live, in the troposphere. Warming is evident over much of the planet with the largest warming occurring over the northern continents. According to the Intergovernmental Panel on Climate Change, averaged over all land and ocean surfaces, the temperature has warmed 1.53 degrees F (0.85 degrees C) from 1880 to 2012. Much of the scientific debate (and political debate, for that matter) is about exactly what is causing this warming. Various natural and man-made factors play roles in climate change. There is natural variation in the solar cycle that impacts our climate and weather. But we as humans also have an impact the weather and climate through such things as increased pollution, increased generation of greenhouse gases, increasing population, and changes in land use. The crux is to determine how big a role each plays and in addition how each one affects the other. Some of the factors that play a role in climate change are: greenhouse gases, aerosols, land surface changes, solar irradiance (i.e., power per unit area produced by the sun), contrails, and volcanoes. Some we have the potential to control, such as greenhouse gases. Others, like volcanic eruptions, we do not.

Most gases in the atmosphere are *not* greenhouse gases. Greenhouse gases are any gaseous compounds in the atmosphere that are capable of absorbing infrared radiation, thus trapping and holding heat in the atmosphere. The greenhouse gases, from strongest to weakest, are as follows:

1. Water vapor: Also the most abundant greenhouse gas.
2. Carbon dioxide (CO_2): Human sources are much less than natural sources, however, carbon dioxide exerts a stronger radiative force than any other human-released greenhouse gas and has a longer lifetime in the atmosphere than many others. "Lifetime" in the atmosphere refers to how long an individual molecule remains in the atmosphere or how long it takes for an increased concentration of a molecule (perturbation) to return to the initial concentration.
3. Methane (CH_4): Second strongest human-released greenhouse gas.
4. Halocarbons (e.g., freons, chlorofluorocarbons [CFCs]): These are strictly man-made gases.
5. Nitrous oxide (N_2O): Yes, laughing gas. The lifetime of this gas in the atmosphere is more than a hundred years.
6. Ozone (O_3): Ozone in the stratosphere is good, ozone in the troposphere (where we live) is bad. The highest concentration of atmospheric ozone is in the stratosphere.

As we can see from this list, some of these greenhouse gases are long-lived and many are man-made. The increasing concentrations of these greenhouse gases, such as carbon dioxide and methane, contribute to the warming trend as they absorb infrared radiation that should be leaving the Earth, but instead, the radiation remains trapped inside, causing a warming, "greenhouse" effect. In addition to causing warming temperatures, many other consequences of these trapped gases are being discovered. For example, as atmospheric carbon dioxide rises, the oceans' chemistry changes and they lose their ability to store as much carbon dioxide, and thus more is released into the atmosphere, creating a vicious cycle. These gases act like a blanket over our atmosphere. The majority of these gases come from the burning of fossil fuels. They also come from agricultural processes (such as the application of fertilizer, burning of crops, and the fact that cows produce methane through their digestive processes), industrial processes (like the burning of fuel for power and heat or from the chemical reactions that occur from the production of iron, cement, and chemicals), and deforestation.

Aerosols are another element in our atmosphere, in addition to these trace gases. Aerosols are either solid or liquid particles suspended in our atmosphere and are usually quite small, on the order of a few microns (μm)—for comparison, the diameter of a human hair is approximately 100 μm. Approximately 90 percent of aerosols are natural (such as dust, sea salt, smoke, and ash), but some are not. Key aerosol groups include sulfates, organic carbon, nitrates, mineral dust, and sea salt. The remaining 10 percent of aerosols are anthropogenic (man-made). These include particles and gases released from things like fossil fuel combustion, biomass burning, power plants, smelters, and incinerators. Though they are small, aerosols have a tremendous impact on climate and on human health, which is affected when we ingest or inhale these particles. The IPCC has deemed that aerosols themselves have an overall cooling effect on climate. Aerosols cool the climate directly by scattering sunlight back into space. These aerosols also impact the climate budget of the planet. In the lower portions of the atmosphere they modify the size of cloud particles, affecting how clouds reflect and absorb sunlight.

When speaking of warming and cooling the planet, no discussion is complete without a mention of the sun and solar intensity. Solar intensity, or the sun's brightness, varies over time for many reasons, but the primary reason for variation is the eleven-year sunspot cycle. The intensity of the sun varies with the number of sunspots that manifest themselves on the surface of the sun. Sunspots appear as dark spots on the sun because they are cooler than the surrounding sun's surface and are regions of highly concentrated magnetic fields. These sunspots follow the aforementioned eleven-year cycle: when there are a lot of sunspots, there is a higher intensity of solar radiation that hits the Earth. A group of scientists ran a global climate computer model that looked at greenhouse gas emissions from 1880 to 1993, combined with solar irradiance, and found that the sensitivity of the climate to solar irradiance is 27 percent higher than its sensitivity to forcing by greenhouse gases as these gases can absorb and emit radiation within the thermal infrared (heat) electromagnetic range. This has led to some politicians using this research to claim that the warming of the troposphere we are currently experiencing is due to "natural" forces alone and not human causes. This is certainly not the case, as it is a combination of this cycle of solar intensity with an increase in greenhouse gases, increasing ozone near the surface, and the ensuing ancillary effects that result in climate change, and we will continue to see increases in the Earth's average temperature and associated consequences, such as reduced ice and snow cover, increased acidity of the oceans, and the raising of sea levels.

We've discussed what is happening in the troposphere, but what about the stratosphere? The stratosphere and its role in weather and climate is still unknown. In addition, the mechanisms linking the stratosphere to the troposphere must be better understood, and we also need better representation of those mechanisms in weather and climate models. The role of water vapor and its transport into the stratosphere, a key role in radiation and chemistry of the models, is also quite uncertain. Atmospheric gravity waves play a large role in climate, but the way they are parameterized (input) in models today is still quite crude. Some of the answers we currently seek are the relation of gravity wave generation to the circulation processes and its effects on climate. But this requires very high-resolution modeling of these events along with a better understanding of the processes that generate these waves and their variations. The lower stratosphere wind and temperature perturbations may also impact tropospheric breaking wave activity. But further study on the interplay between these two levels of our atmosphere remains to be done, and it is my personal hope that continued glider launches, like the Perlan, will be successful, so that we can continue to learn more about this final frontier of atmospheric science.

The bottom line is that although there are complex relationships contributing to climate change, some man-made, some naturally occurring, we are armed with several very important facts. We know that the global average surface temperature is rising, the temperature in the troposphere is rising, the stratosphere is cooling, the ocean is warming, sea levels are rising, oceans are becoming more acidic, and ice caps are melting. These effects in turn are causing changes in our weather patterns in many locations around the world. We also know that humans are causing some of this, and *this* is where we as a global society can do something about it, such as reducing our carbon footprint, recycling, and using less water. I was reading our local town's newspaper and I came upon the following letter to the editor.

Thursday, January 22, 2015
North Lake Tahoe Bonanza

Letter to the Editor

Are we afraid to talk about climate change?

As I sit on the Diamond Peak chairlift staring down at the dirt, hike to the top of Tunnel Creek in a T-shirt, and see temperatures forecast in the 50s for Lake Tahoe this week, I am feeling more and more anxious.

I silently say to myself: climate change. I whisper it under my breath: climate change. I sheepishly, quietly say it to my friends: climate change—and then just as quickly move on to the next topic.

Why am I avoiding talking about climate change? Why am I making excuses and telling stories like La Niña, El Niño, high-pressure systems and low-pressure systems.

I know what scientists have warned as the climate debate destabilizes: increasingly warm winters, drought, and significant lack of precipitation.

The extremes of a geographical area will become the new norm (worse drought in the West, stronger hurricanes in the Gulf, flooding in coastal areas).

So why am I avoiding talking about climate change? In my brain's simplest deduction: It's scary, it's big, and I am afraid to accept that it is real.

But unless I get real about climate change—unless we all get real about it—the problem will only worsen. It is time for honesty, communication and truth. And action.

I hope for an epic February. I hope for feet of powder in March. And while I continue to wax my skis and hope, I also need to get honest about what is really going on.

Amy Guinan
Incline Village

I read this and realized that though some of the statements may not be scientifically accurate, such as the expectation for stronger hurricanes, the overall gist of Amy Guinan's brave letter was very real. The fact is that many of us do not know what to think about climate change—what is it, what is causing it, and what we can do about it—and that makes it scary. It is like the child who thinks there is a monster under the bed and goes to bed scared every night. It is better to get out a flashlight and look under the bed. But what makes this so difficult? Misinformation! This can come in many forms, whether it is articles not based on any science to political rhetoric.

A local college paper recently published an article written by one of its students in which she made the following statement: "The atmosphere is polluted with greenhouse gases." It brought home to me how many people misunderstand what greenhouse gases really are and what they do. She thinks that all greenhouse gases are pollutants. Not true. Don't forget, the most abundant greenhouse gas is water vapor! But there are still grains of truth in this well-meaning, if misinformed, article.

I have heard in the media many uninformed and preposterous pronouncements regarding the rather complicated scientific issue of climate change—to wit, one particularly glaring example of a classic *reductio ad absurdum* argument in which one of our illustrious, but very poorly informed, members of Congress, U.S. Senator James Inhofe, stood on the Senate floor holding up a snowball and declared this as evidence that global warming doesn't exist, and the whole thing is a farce. Socrates could have added this one to his quiver of examples of ridiculously flawed reasoning. It's just a tiny bit more complicated than that, Senator. It is a good reminder that correlation does not equal causation. Though the increase in the incidence of autism is correlated to the increase of organic food sales, that does not mean or prove causation, that organic food causes autism. The Senator's sophomoric logic—that because there is snow, somehow climate change cannot exist—reminded me of the a cartoon I recently saw showing the *Titanic* sinking, with the back part of the ship down in the water but the nose of the ship high up in the air. The caption read, "The ship can't be sinking. My end just rose 200 feet." It brings home the problem of not understanding that this is a *global* issue and not one limited to one location or nation during one particular time or season. China's top atmospheric scientists have finally acknowledged that climate change exists and stated that "as the world warms, risks of climate change and climate disasters to China could become more grave." China is the world's largest emitter of the greenhouse gases that contribute to climate change, including carbon dioxide and methane. The chief of China's Meteorological Administration, Zheng Guogang, stated that their temperatures on mainland China have been higher than global averages, and that they need to be proactive about their emissions in order to keep things from escalating even further.

But climate change is a global issue and we cannot fixate on one specific country or industry. Every country is an emitter of greenhouse gases to some degree and every country is impacted by emissions from countries around the globe. My husband, Alan, and I were landing in Pago Pago (pronounced "Pango Pango"), the capital of American Samoa, with Steve Fossett and a couple of the Perlan team members in 2003. We were traveling in Steve's Cessna Citation X jet from Hawaii to New Zealand and needed to refuel. American Samoa is composed of five islands (and two coral atolls) and is an unincorporated territory of the United States. The islands are located in the heart of Polynesia in the Pacific Ocean. When we arrived, others on the team who had been there before told us how lucky we were that "they were not burning." Apparently one day a week is a day is set aside for the burning of garbage, and on that day the air becomes so thick with smoke that one can hardly breathe. This is their *best* solution to taking care of their garbage, as opposed to the high expense of shipping it to a mainland somewhere else. Their seventy-six square miles of land total is not capable of handling the garbage if it was buried over time. But since these people are literally out in the middle of the Pacific, the main and immediate impact from the unhealthy fumes is on the inhabitants of the island themselves, as well as any visitors. However, as we now know, some of the pollutants spewed into the air can make it vast distances across the globe and impact other locations and peoples. This is just one example of how no decision regarding the climate is cut and dried, and what occurs in one location on the planet affects the rest of us, too.

As it turns out, there maybe *is* a monster under our bed. But if we take the time and have the courage to look, then we will realize that this monster is something that we can study and understand—and that we can do something about it if we act now.

FIFTEEN

Plane Crashes

> *If Beethoven had been killed in a plane crash at the age of 22, it would have changed the history of music, and of aviation.*
>
> —Tom Stoppard

The evening of July 16, 1999, was not a good night for a novice pilot to be flying to Martha's Vineyard.

"Traffic! Traffic!" the traffic avoidance/collision system blurted out to American Airlines Fokker 100 commuter commercial Flight 1484, which was on approach to Westchester County Airport, White Plains, New York.

> At 2052:22, the controller: "American fourteen eighty-four, traffic one o'clock and five miles eastbound, two thousand four hundred, unverified, appears to be climbing."

2052:29, Flight 1484: "American fourteen eighty-four, we're looking."

2052:56, the controller: "Fourteen eighty-four. Traffic one o'clock and, uh, three miles, twenty eight hundred now, unverified."

2053:02, Flight 1484: "Um, yes, we have uh (unintelligible). I think we have him here [on our radar]."

2053:10, Flight 1484: "I understand he's not in contact with you [the air traffic control tower] or anybody else."

2053:14, the controller: "Uh, nope, not talking to anybody."

2053:27, Flight 1484: "[This plane] seems to be climbing through, uh, thirty-one hundred [feet] now. We just got a traffic advisory here."

2053:35, the controller: "Uh, that's what it looks like."

2053:59, Flight 1484: "Uh, we just had a [unintelligible]."

2054:12, the controller: "American fourteen eighty-four you can contact tower nineteen seven."

2054:15, Flight 1484: "Nineteen seven, uh, we had a resolution advisory, seemed to be a single-engine Piper Comanche or something."

2054:21, the controller: "Roger."

That was the actual conversation between an air traffic controller and the crew of American Airlines Flight 1484 on the night of July 16, 1999, as they narrowly averted a midair collision with an oblivious pilot in a Piper Saratoga. The American Airlines pilots diverted and avoided colliding with the small plane, which, until they diverted, was heading directly toward them, and landed safely. The passengers were unaware of the potentially deadly situation. The Piper Saratoga was another story.

The Saratoga continued flying northeast up the Connecticut coastline, cutting across the Rhode Island Sound headed over to Martha's Vineyard. From there, it was scheduled to go on to Hyannis, Massachusetts. But before reaching the Martha's Vineyard Airport, it crashed into the Rhode Island Sound. The Saratoga's inexperienced pilot was flying without his certified flight instructor, who accompanied him on many occasions and who had offered to fly along on the fateful July evening. Another Saratoga pilot destined for the same airport at Martha's Vineyard had cancelled his flight from Essex County Airport, due to poor weather that would have been especially difficult for such a small plane piloted by a terribly inexperienced pilot.

What was the weather like that night? The skies were clear and the visibilities along the flight route were three to five miles, which was passable for flying, but there was still haze and mist. And it was nighttime. That's not a problem for a seasoned, instrument-rated pilot. But the pilot of the Piper Saratoga was not instrument-rated and had a total of just 310 hours of flight time, only 55 of which were at night, and 72 of the 310 were flown with a flight instructor on board. Experience, however, is measured not just by the amount of *time* flying but by the *type* of aircraft flown, in what's referred to as "time in type." One pilot may have 1,000 hours of daytime flying, all with clear skies and in a small "putt-putt" aircraft; another may have 1,000 hours of daytime, nighttime, and bad weather flying in a high-performance jet. Same number of hours, technically, but there is quite a difference, and this is what pilots mean when they say "time in type" (which is also a requirement for the insurance companies that insure these pilots). Putting this into perspective, Chesley "Sully" Sullenberger had just under 20,000 hours of flying time and his first officer, Jeffrey Skiles, had approximately 16,000 hours of flying time when they landed their Airbus A320 on the Hudson River on January 15, 2009, a feat now known as "The Miracle on the Hudson." Regarding this miracle, Captain Sullenberger stated, "One way of looking at this might be that for forty-two years, I've been making small, regular deposits in this bank of experience, education, and training. And on January fifteenth the balance was sufficient so that I could make a very large withdrawal." He not only had a lot of hours, but his time-in-type hours were vast and encompassed a range of experience: day, night, and in a myriad of weather conditions in a plane comparable to the craft he was piloting that day.

So what happened to the Saratoga? As a forensic meteorologist, I'm inclined to start by analyzing the atmospheric conditions, including the lighting conditions. There are weather stations all over that region of the northeast coast of the United States. Many airports, in fact, have automated weather stations, which record and report weather conditions such as visibility, cloud cover, pressure, and temperature. As the American Airlines pilot stated, the weather was "poor" that night, even though most laymen would not have described that way: the temperatures were in the lower to mid-seventies with mostly light winds, mostly clear skies, and relative humidities in the 60 percent to 80 percent range. Sounds like a lovely evening . . . unless you are flying using visual flight rules (VFR)—that is, not really using the instruments in the aircraft but mostly relying on visual cues outside your windshield. The difficulty with this came in to play that evening for the Piper Saratoga, as the weather stations along the flight route from Essex County Airport to Martha's Vineyard to Hyannis reported reduced visibilities of three to eight miles, with mist or haze. Haze is when particles suspended in the air reduce visibility; mist is when airborne microscopic water droplets reduce it. The difference between the air temperature and the dew point temperature—the temperature to which the air must cool for dew to form—is used to distinguish between haze and mist. On the night of July 16, 1999, both mist and haze were reported, a "perfect storm" of conditions for a pilot with low flight time and overall inexperience. It was not seemingly terrible, but just bad enough to pose a threat to an unseasoned pilot such as the one piloting this Piper Saratoga. And because nothing happens in the sky in isolation, this pilot who was out of his depth then became a threat to others in the air around him, like the nearly one hundred people on American Airlines Flight 1484.

For a pilot flying over a body of water that has no lights on it, or clear visual references, haze and mist combined with nighttime conditions can, and often do, as in this case, become a recipe for spatial disorientation and the inevitable ensuing disaster. As this aircraft flew over the Rhode Island Sound toward Martha's Vineyard, the moon was setting in the west, just barely above the horizon and only 19 percent illuminated (it was a waxing crescent, for you astronomy buffs). Sunset for this location occurred at 8:14 P.M., and the sky was so faintly illuminated that it was practically imperceptible while flying over the water. From the time that the Piper Saratoga departed Essex County Airport until the time it crashed, there is no record of its pilot communicating with any flight service station or air traffic controller. Even though he was not on a VFR (visual flight rules) or IFR (instrument flight rules) flight plan, he could have, for example, asked air traffic control for what's called "flight following," which is a service offered by ATC if they aren't too busy with IFR traffic, to follow a flight by giving the pilot a discrete transponder code and monitoring the aircraft and supplying advisories along the route of flight. (IFR—instrument flight rules—require that the pilot file a flight plan, be an instrument-rated pilot, and that the aircraft be certified for instrument flight.) As inexperienced as he was, the pilot may not have known that the flight following service was available to him—it might have saved the three lives of the pilot and his passengers. Many low-time, inexperienced pilots are either not aware that flight following is available or more likely afraid to be obliged to follow ATC instructions that may be confusing to them. Radio procedures and ATC instructions are extremely intimidating to new pilots, especially in high-traffic areas. Even though a primary instructor may have told them they could avail themselves of this service, new pilots are not usually very comfortable doing so. In addition, the service is only offered if the ATC controller isn't busy, and that is not too often these days. Very often inexperienced pilots are quite intimidated by the rapid communications that come over the radio from ATC. There was not even a distress call from the aircraft on this fateful evening.

The crash's probable cause, as determined by the U.S. National Transportation Safety Board (NTSB) was "spatial disorientation." Humans perceive their spatial orientation through visual perception (with the eyes), vestibular perception (with organs of equilibrium in the inner ear), and proprioception (with receptors in the skin, joints, muscles, and tendons). Experienced pilots know and often tell inexperienced pilots, "You cannot trust your inner ear." Anyone who has been skiing in whiteout conditions, driving in thick fog, or trying to find the surface of the ocean after being clobbered by a wave has experienced the validity of this statement. This holds especially true for flying aircraft, as one is now in a three-dimensional environment. A very important part of aircraft instrument training is learning to deal with the spatial illusions and false indications one encounters in flight and to fully trust the instruments (including cross-checking on their reliability, as well, as they too fail on occasion). According to the Federal Aviation Administration (FAA), spatial disorientation as a result of continued visual flight into adverse weather conditions is statistically one of the leading factors in fatal aircraft accidents. On top of that, 90 percent of the general aviation accidents attributable to spatial disorientation are fatal.

During the final moments of this Piper Saratoga's fateful flight, when the plane was only seven miles from the shoreline of Martha's Vineyard, it made a right turn, then a descending left turn, and then an ascent up to about 2,600 feet above the water, and then another descent. It then made another right turn and accelerated from about 900 feet per minute to over 4,700 feet per minute (the maximum safe descent speed for that aircraft was 1,500 feet per minute). It struck the ocean nose-down. The flight path, and the type of vertical impact that followed, was clearly indicative of the spatial disorientation of the pilot. The sad part is that the aircraft had a working autopilot which, if programmed properly and engaged, could have flown quite safely over the Rhode Island Sound to its destination airport. However, as is also often the case, newer pilots either distrust the autopilots or are not comfortable programming them. In the case of this aircraft, the programming would have been relatively easy—the trust and instrument scanning to make sure the autopilot is doing what it's supposed to do is perhaps more problematic for inexperienced pilots. The result is that "newbie" pilots don't engage the autopilot; they try to fly while alternating between looking at the instruments and outside the aircraft, hoping they will find some sort of reference to the ground. This greatly exacerbates the rate of disorientation they experience. Combine this with not checking in with air traffic control, and the disaster could have been compounded a hundred times.

Who *was* the pilot of the Saratoga? It was none other than John F. Kennedy, Jr., the son of the late President John F. Kennedy. The other two fatalities on the plane were JFK Jr.'s wife, Carolyn Bessette, and his sister-in-law, Lauren Bessette. The truth is, aside from the famous names, what happened to the Piper Saratoga is all too common. The famous, the wealthy, and the "average Joe" alike tend to overestimate their abilities and underestimate the weather conditions. Fear and stress also play a role. I am reminded of a comment by Wilbur Wright: "It's possible to fly without motors, but not without knowledge and skill." Flying is similar to being a surgeon: it takes a lot of study and one cannot do it in his or her spare time. If you don't do it all the time, you're not going to be very good at it. To further that point, ego is no substitute for experience and good judgment about one's own abilities. Money and ego are not going to help without knowing your own limitations. The lack of good judgment, as in this case, has, sadly, cost the lives of many trusting passengers as well.

After any of these tragedies there inevitably follow hoards of lawyers filing suits.

The door to the courtroom opened loudly and a bailiff stepped out. "Dr. Austin to the stand," he uttered, stridently. I was off and running on another case testifying about the weather conditions that prevailed during the accident. Learning the ropes of how things work in a courtroom took a good deal of time for me and has served me well on the road to becoming an effective expert witness. I will unabashedly say it is not an easy status to obtain, by any means, and this case, given the people involved and the media frenzy surrounding it, was going to be especially dramatic.

As I was the only one waiting on the absurdly uncomfortable bench outside the courtroom, the bailiff had no choice but to assume that I was Dr. Austin.

"Thank you," I replied, gathering the notebook containing all of my work on the case. Walking past the gallery, I focused on the layout of the courtroom. As I walked past the plaintiffs and defense attorneys at their respective tables, I glanced over at the jury and scanned for the key members—the specific jurors who needed to understand my scientific testimony, so that they could explain it to the others if necessary. In this case, they were the fireman, the engineer, and the private pilot. I saw the foreman. I also saw the lady that smiles at everyone no matter what they have to say and the man who falls asleep after the lunch break. I knew who sat where. It's funny how people are such creatures of habit. Even when jurors are not assigned seats, after a few days at trial they tend to keep their same seats day after day.

In many high-profile cases, expert witnesses know the occupations of all of the jurors, the habits and personality of the judge, and the cross-examination techniques of the opposing attorneys. Sometimes it is the law firm that educates the expert in these areas and other times the legal team hires a company that specializes in preparing expert witnesses for specific trials. "If the judge leans back in his chair and rubs his forehead, he is bored" and "If the judge starts scratching the back of her neck at the hairline, she's contemplating your testimony and not completely understanding it" are a couple of tips I have received regarding reading judges.

For some cases there are even "mock jury trials" prior to the actual trial. These are held by one side only, usually without the knowledge of the other side. They are done to learn what works, what doesn't, what evidence the jury will find persuasive, and where the case might have weaknesses. Mock trials are done with the expert witness's real data and information, but the testimony is generally presented by the attorneys, not the experts themselves.

I once testified in a case that had a "shadow jury," which is a bunch of paid regular citizens who come in, sit, and listen, and report their findings to a consultant hired by one of the litigants. The other side usually discovers the "phantom jurors" within the first day or two, depending on the size of the audience, and then promptly complains to the judge. Usually they'll demand the presence of the phantom jurors be disclosed to the real jurors.

"Reporter, please swear in the witness," the judge announced as I neared the witness stand.

I stopped walking, standing in my high heels, holding on to my large and, as always, very unwieldy notebook, and turned to the court reporter.

"Please raise your right hand," stated the reporter.

I raised my right hand and made sure I focused on the court reporter's eyes while she said, "Do you swear to tell the truth, the whole truth, and nothing but the truth?"

I paused, as I always do, before answering this question. I never know if the reporter is going to follow the statement up with "so help you God." About 70 percent of the time they do not, but you never know. In certain states, like Texas, the odds are about 90 percent that they will follow it up with "so help you God." On that day she did not.

"I do," I answered, with conviction.

"You may take the stand," stated the judge.

I walked up and placed my notebook on the witness stand, sat down, and adjusted the chair and the microphone to suit my height. Once I was settled, the judge, without even looking at me, blurted out, "Please state your full name for the court."

"Elizabeth Jean Austin," I replied, pronouncing each name separately and clearly. I was at yet another trial testifying about the atmospheric conditions surrounding yet another deadly plane crash.

Even if they don't admit it, most people are inherently afraid of flying. Some of this fear is justified, but most is not. The National Safety Council statistics show that one's odds of dying in a car accident during his or her lifetime is 1 in 98, whereas the odds of dying in a plane crash (commercial, commuter, or private) is just 1 in 7,178. Statistically, flying is much safer than driving. That being said, if possible, I do try to plan my flight routes around the weather, depending on the time of year it is, my destination, and my work schedule, and with the help of my husband, Alan, who was a professional pilot, I also pay careful attention to the aircraft type. Is this action warranted? I think so. Why? About 22 percent of all plane crashes are weather related and 31 percent of all Part 135 (commuter and charter flights) crashes are weather related. The major airlines operate under Part 121 (this covers all commercial airlines), and are subject to a quite different set of rules and procedures—preferable ones, in my opinion, as they are much more stringent. For example, Part 121 operators are required to send their pilots for recurrent training every six months, which entails both written and flight-simulator proficiency testing. (There are other particulars regarding operational guidelines of Part 121 that are not pertinent to this discussion.) The Part 135 (air taxi) section of the FAA regulations is less stringent; it governs many charter operators and some commuter flights. Most passengers would really have no way of knowing under which set of rules their carrier operates, except that if you are on a large aircraft, you can be sure it's a Part 121 operation (this obviously applies to U.S. carriers). The FAA defines a large aircraft as any aircraft with a certified takeoff weight of more than 12,500 pounds. That definition however covers many aircraft, like private jets, that don't operate under Part 121 rules. For the sake of the discussion here, when I use the term large aircraft I'm speaking of commercial airliners that operate under the flagship names—American, United, etc. If you're flying a commuter plane, it could be operating under either Part 121 or Part 135. Strictly speaking, commuter airlines may have a version of the flag carriers logo on the side of the plane, but they are operated by completely separate and different entities, a fact that I would assume most of the flying public does not know. The most stringent, and in my opinion, safest rules are the Part 121 regulations that the flag carriers are obliged to operate under.

The statistics are interesting and also telling. Focusing on commercial airlines and flights, the safest aircraft in terms of fewest fatalities, hands down, is the Boeing 777. Except for the imbecilic pilots of Asiana Airlines Flight 214 from South Korea to San Francisco, back in 2013, which crashed on final approach to the airport because the aircraft was too low and hit short of the runway, killing three people, there have been no fatal accidents to date in a "Triple 7." The second safest aircraft is the current generation of the Boeing 737 (the older generations—737-500, 737-400, and 737-300—hold the number four spot). Number three is the Airbus A320 series (France), though I'm not sure how the most recent crash in the French Alps will affect this statistic; and coming in at number five is the Canada Air Regional Jet (Canada). Thankfully, all of these aircraft are hugely popular with commercial airlines today. The four aircraft that have to date been flown fewer than two million miles without being involved in a fatal crash are the Airbus A340 and A380, and the Embraer 170 and 190 (Brazil).

Of course, I cannot list the top aircraft for safety without also listing the least safe ones. Number one on that dubious list—the most unsafe—is the LET-410—a twin-engine short-range transport aircraft (Czech). Next are the Antonov AN-12, a turboprop transport aircraft (Russia); the Ilyushin Il-76, a four-engine airliner (Russia); and the CASA C-212, a turboprop medium-transport aircraft (Spain).

For a time, when Alan was based in San Juan, Puerto Rico, he flew the CASA C-212 for American (Eagle) Airlines. His comments were that an ugly aircraft usually flies ugly and the CASA is one of the ugliest he'd ever seen. The recruiter, of course, had told him he'd be flying the ATR-42 and ATR-72, which all the American Airlines Caribbean posters featured—because they knew that if they'd shown the CASA, they wouldn't have gotten one recruit to sign up to fly that heap. The CASA had many ugly traits, but at least one very ugly one cost several lives in two separate crashes—amazingly, at the very same airport. During Alan's training, the unlucky captain assigned to train new pilots on the CASA cautioned the crews to be very careful about bringing the throttles back too far when descending. If they did, something—which the maintenance people had obviously not solved—would cause the engines to go into "beta." On turboprop aircraft, "beta" is the mode during which the propellers go into reverse pitch; it is supposed to be used *after* landing to aid in stopping the plane by providing braking action. When this happens in the air while descending, however, the aircraft will suddenly become uncontrollable.

Mayaguez airport is at the opposite end of the island of Puerto Rico from San Juan. There are a fair amount of business interests there and it was a regular daily stop for American's CASA routes. It sits on the coast facing the waters between the Dominican Republic and Puerto Rico. The rarely used instrument approach (as opposed to the visual approach) required the pilot to fly past the airport out over the strait between the two islands, out over the water, and make a 180-degree descending turn to come back and land. (The strait was named "shark alley" by the locals because so many of the people who try to escape the Dominican Republic by attempting the swim don't make past the sharks.) The visual approach is very similar, but is accomplished by descending through an even sharper 180-degree turn, closer in, over the shoreline, and then lining up for landing. The two aforementioned CASA crashes in Mayaguez happened in very similar ways. In the first, the pilot pulled the throttles back in the turn while descending at a fairly rapid rate, and one of the engines went into beta mode. At that altitude, there was absolutely no hope for recovery. The plane impacted the ground at too high a speed, caught fire, and all the souls on board perished.

Alan said that with full flaps selected, the CASA comes down like a grand piano even with both engines operating normally.

Not too many years later, another CASA made the same visual approach. A very young and inexperienced new hire was flying his first leg to Mayaguez. The captain, a friend of Alan's, and a seasoned flyer from St. Thomas, was monitoring the approach. Once again, just as the plane rolled out of the turn, the number one engine went into beta. Alan said that listening to the last transmission over the radio from the young pilot was horrific. "I didn't do anything!" yelled the young copilot, seconds before he perished. The plane impacted, burned, and all were again lost. American Airlines has since discontinued service to Eugenio Maria de Hostos Airport at Mayaguez, and no longer flies the CASA. (To the best of my knowledge, no American carriers currently use the aircraft.)

The main airport in Puerto Rico is Luis Muñoz Marín International in San Juan. Alan said that many of the passengers who arrived at Luis Muñoz Marín from abroad expected to be delivered to their island getaways via the same type of aircraft they arrived in; namely a Boeing 757, or at least something comparable to that. But much to their chagrin, they found CASAs lined up like so many ugly ducklings along the tarmac between the main terminals. Often during the daily downpours, when the gate agent escorted the passengers out to board one of the beasts, Alan heard the disgruntled passengers say, "Are you kidding? I'm not getting on that thing!" Many actually refused to board. They were the smart ones.

North America has the lowest rate of fatal aviation accidents of all regions on the globe. The rate of fatal air accidents in Africa, however, is more than seven times that of all the other regions combined. Not too far behind Africa is the non–European Union region of Europe, then South America, and then the Middle East. Statistically, the safest months to fly are September and October. By far the most dangerous in terms of fatal accidents are December and January, then February, April, and May. There are three hours during the day (all local time) during which most fatal accidents occur: number one is from six P.M. to seven P.M.; the others, in order, are ten A.M. to eleven A.M., and eight P.M. to nine P.M.

In terms of airlines, Dustin Hoffman's character in the movie *Rain Man* had it correct:

Charlie: "Ray, all airlines have crashed at one time or another, that doesn't mean that they are not safe."

Raymond: "Qantas. Qantas never crashed."

Charlie: "Qantas?"

Raymond: "Never crashed."

Charlie: "Oh, that's gonna do me a lot of good because Qantas doesn't fly to Los Angeles out of Cincinnati, you have to get to Melbourne! Melbourne, Australia, in order to get the plane that flies to Los Angeles!"

Every year, lists of the safest, and least safe, airlines are announced from a few sources, including the U.S. National Aviation Safety Data Analysis Center and the National Transportation Safety Board. One enduring paragon of safety is indeed Qantas. Qantas (the Australian airline) is still ranked as the safest airline. They use cutting-edge technologies for navigation and precision approaches. Also, they monitor every aircraft in their entire fleet in real time, by using satellite communications. That means they are able to detect problems before they become major safety issues.*

The next safest airlines, after Qantas, are: Air New Zealand, British Airways, Cathay Pacific Airways, and Emirates. No U.S. airlines make the top ten safest, but this is perhaps due to the fact that non-fatal accidents are included in the rankings. Many U.S. airlines, such as Southwest Airlines and JetBlue, have had zero fatalities in their operating history.

There are so many methods for ranking airlines, but the ones that I care about most are safety and fatalities. The listing of the world's most dangerous airlines, like the listing of the safest, varies from year to year,

* An interesting side note regarding rotorcraft safety: ten out of every one hundred helicopter pilots are women but they account for only three out of every one hundred accidents (John King, *Flying Magazine*).

but airlines that are on this infamous list usually remain on it year after year. The top five most dangerous airlines, starting with the worst are: (1) Cubana Airlines, (2) China Airlines, (3) Iran Air, (4) Philippine Airlines, and (5) Kenya Airways.

Most aviation accidents, both commercial and private, occur during the takeoff or landing phase of flight, not during cruise flight. A rare few are even caused by pilot suicide, and, unfortunately, we just had to add one more to that sad list: Germanwings Flight 4U9525. I have been asked to investigate some of these possible suicides by looking at the winds along the final portion of the flight's path. In some cases the pilots had become separated from the aircraft—meaning they jumped off the plane—and we had to determine if the aircraft fell to the ground on its own or if it had been steered toward the ground on purpose. That meant I had to use the weather conditions and physics to calculate where a person's body would end up if they deliberately jumped from the aircraft. The Germanwings plane apparently was in the "deliberate" category, meaning he steered the flight into the mountainside himself.

Weather-related accidents comprise over 21 percent of all aviation accidents each year. Of these weather-related accidents in the U.S., the huge majority (over 86 percent) are on Part 91 flights, which are the general operating rules/regulations for private pilots on private aircraft. Part 135 flights—charter and some commuter flights—come in second at a little over 6 percent. Interestingly, Part 137 flights, agricultural aircraft operations come in third.

I'm not a pilot, but on a side note I, and I'm sure many other "civilians," find it interesting that the training, recurrency, which are the medical requirements and weather minimums for Part 121 airline pilots, are quite strict and rigorously enforced by the airlines and the FAA. The average Joe who often can afford quite a sophisticated aircraft is allowed to fly under much, much less stringent rules, especially in regard to weather minimum requirements. This section of the FAA regulations, as I said above, is referred to as Part 91 and is less stringent than either Part 121 or Part 135, as previously discussed. So you have a situation where pilots who fly infrequently, as a Part 91 pilot, can very easily get themselves into a lot of trouble very quickly. I see this over and over, usually with tragic results for the pilot and the oblivious passengers whose lives he takes. My case files are full of these types of accidents.

The worst thing about weather-related accidents is that 25 percent of them are fatal. The NTSB has highlighted weather on its Most Wanted List. But I must put these facts into perspective. I have long realized that, though weather may play a role in numerous plane crashes, it's rarely named as the main cause of the accident. For instance, the NTSB may conclude that the probable cause of an accident was the pilot's poor judgment or pilot error—in other words, the pilot should not have flown into the weather they encountered. Though weather, be it wind shear from a thunderstorm, icing in clouds, or poor visibility, may not be named as the primary cause of the crash, it still plays a significant a role, hence the importance of pilots learning to fly in all types of conditions, and developing the judgment of when not to fly at all, depending on their particular aircraft.

The number one culprit in weather-related accidents is wind. Wind is a factor in 48 percent of these accidents. It can cause crashes due to very strong winds or wind shear (changes in wind speed and/or direction over a relatively short distance). This is followed by poor visibility (20 percent), as we saw with the disorientation that caused JFK Jr.'s Piper Saratoga crash; low cloud ceilings (the cloud bases); and then turbulence in third at just over 9 percent, which poses threats to aircraft such as causing in-flight breakups. So even though most of us detest turbulence, it is not the main culprit in most weather-related plane crashes. I will tell myself this on my next flight so when it gets rather bumpy I can sit back and relax . . . perhaps after two glasses of wine!

SIXTEEN

A Canary in the Coal Mine

I find hope in the darkest of days, and focus in the brightest. I do not judge the universe.

—The Dalai Lama

There it was—the wonderful old red gas pump in front of the Silver City General Store that marked the gateway to Mineral King Valley, which ran two miles farther up the winding dirt road in the Sequoia National Park. This part of the Sierra Nevada Mountains is one of my favorites and I have been there often. To me that gas pump in front of the old store was like a beacon for this old mining town. It always gave me goose bumps for the excitement of things to come. It was at that store where I would stock up on anything I might have forgotten to pack for my adventure. The drive from California's Central Valley up to the Mineral King Valley takes you up a long, winding, and always windy, road that becomes mostly dirt once you pass the town of Three Rivers. Three Rivers lies at 875 feet elevation and the Mineral King Valley is at about 7,500 feet. Being a glacial valley, Mineral King has some of the most beautiful views in every direction.

It was midsummer and the Silver City Store was buzzing with people stocking up for their day hikes or extended backpacking trips. A couple of months before, I had talked some of my friends into this trip, which we painstakingly planned. Since we were all high school girls at the time, I decided to ask my climbing club if any other adults would care to join us. Two married men whose children had known each other since childhood volunteered to join us. But before we committed, a couple of my friends and I went to one of these volunteers' houses and met the family just to be sure that all was safe. I could already envision the headlines—"Santa Monica High School Girls Raped, Tortured and Murdered in the Sierra"—but my fears were unfounded. These two lovely men drove a separate vehicle up to meet us at the Mineral King ranger station. With all our gear, my girlfriends and I piled into my beloved 1980 Ford Bronco, and I drove us there. We arrived late, not the best time to get those "headlines" out of my head. We set up camp in and around the Bronco and waited for dusk.

Even though I was only seventeen years old, I felt seasoned enough to lead a packing trip, as I had been backpacking in the Sierra, the Rockies, and the Cascades since the age of twelve. In the five years since my first trip, I had learned how to tie some important knots (thanks to the wonderful *Ashley Book of Knots*), as well as how to use a compass, ropes, an ice axe, and crampons. I had passed the Red Cross's Advanced First Aid and CPR classes and had become a member of the Red Cross Youth Disaster Action Team. I had even taken mountaineering classes and had learned how to tie my own fishing flies. This came in handy later, as I matured and took longer and longer backcountry trips, for which it was impossible to carry a full supply of food. I had to carefully plan my routes to cross streams and lakes at key locations, where I caught fish with my pre-tied flies. The mountaineering classes involved book work, written exams, field trips, and field tests, such as rappelling down cliffs, proper knot tying, and self-arresting (stopping oneself from sliding or falling) down an icy mountainside with an ice axe. Little did I realize back then that these skills would come to be important for my future field work and backcountry adventures as a scientist. I had packing down to a minimalist art form. But even so, I erred on the conservative side for this trip and packed more than I usually did, since my friends were with me and I knew how quickly things could turn from glorious to horrible.

Two years earlier, I had been with a group on a two-week backcountry trip. It was our first day, and we were hiking out of Mineral King up to Sawtooth Pass. The ten-mile hike takes you from approximately 7,500 feet elevation to almost 12,000 feet. The final 2,000 feet of vertical elevation gain to the top of the pass covers only about a half a mile. But what makes it even more arduous is that the pass faces west, meaning it gets sun most of the day and doesn't have a tree or even a shrub on the final portion of the pass. So there is no shade, and we were hiking on decomposed granite, which is basically sand. Nearing the final mile to the top of the pass, hiking on the steepest portion now, I began taking steps up the sandy pass; for each step up my foot slipped three-quarters of the step backward. I continued on this way. I was so thirsty, but the thought of stopping and taking off my heavy pack on the steep incline to dig out my bottle of water was too much to bear. The first day is always the worst. Packs were heavy with all of the gear and I was not in hiking shape. The pack dug painfully into my waist and shoulders. There was no hiker near to get my water out of my pack without my taking it off, so I just continued on and on. The top of the pass seemed like it never got any closer. Just as I finally thought I was nearing the top, I realized, unfortunately, that there was still a lot more trail in front of me.

After what seemed an eternity, I finally reached the top of Sawtooth Pass. The view was breathtaking and my thoughts about dying of thirst were quenched by the incredible feeling of euphoria and accomplishment. Looking down toward the east, I saw my destination for the night, Columbine Lake. The lake is at 10,800 feet, but from my vantage point at the top of the pass it felt like I could reach out and touch it. I scanned over to see Black Rock Pass, where the rocks are so dark against the lighter-colored mountains that they seem fake. After a brief rest and some much-needed water, I headed down the trail, which wound around to a small lake called Cyclamen. I peeled off the trail I was on part of the way down toward Columbine Lake. The route I had to take led me down some ferociously steep and rocky faces that eventually would morph into gigantic boulder fields, so I tried to detour slightly for an easier route. I jumped downhill from one boulder to another; the substantial weight of my backpack launched me forward so much that my landings were a little more perilous than I would have liked. While eager to get to the lake, I had to concentrate on every jump so as not to crash. Many of the boulders had nasty jagged edges—some small, some the size of Volkswagens—and many were also dangerously wobbly. It took most of the *oomph* I had left to plan each landing spot so as not to fall. I was so close to the lake, and the orange I had in my backpack was calling to me. I was thirsty and needed an energy boost.

My orange, however, was the wrong thing to be focusing on at that moment. I was about to prove that losing focus in precarious situations can be dangerous. I should have been focusing on the rocks, because I landed on one that didn't have a distinct flat spot. I tried to balance myself and leap over another, knife-edged rock so as to get to the rock beyond it. It was a terribly unsuccessful attempt. My foot slipped on the boulder as I jumped off and I tumbled downhill, trying to do everything I could to save myself. The weight of my backpack didn't help; it just pushed me harder down the hill. I went tumbling down and crushed my shins directly into a huge knife-edged rock and ended up pinned against a boulder farther down the hill.

After the initial shock, I did a damage check and was relieved to see that I did not have any broken bones, nor did I crush my skull. A miracle! Then the pain hit. My shins were on fire. While still lying on my side with my pack on, I looked down at my legs and was literally horrified by the sight. Raw flesh *and bone*! I had gashes that were so deep and long I could see my shinbones. Immediately I felt faint. Not much bleeding, but then there isn't much flesh on the shins. I did notice, however, that I was bleeding from just below my right knee—an ugly deep gash below my kneecap. Trying not to focus on where I had to go, or on my shins, which looked like hamburger, I carefully worked my way on down to the lakeshore, which, luckily, wasn't much farther and, fortunately, didn't have any more nasty rocks to get me. The pain was incredible, but I finally made it to a dirt path that went around part of the lake. Luckily, my fellow hikers began to arrive and saw immediately that *something* unpleasant had happened to me. There were a lot of exclamations, like, "Oh, my god, you are beet red, what happened?" "Look at your legs, oh, I can't look!" and "Are you okay?"

The looks on their faces weren't very helpful, but the contents of their first aid kits were. At first I was pleased that my shins didn't bleed much, but I soon became concerned about the risk of infection. I carefully cleaned my wounds and held my breath as I dabbed the reddish-brown Mercurochrome, which is a disinfectant, into the mess I'd made. I'd wear that reddish brown on my legs like a badge for the next two weeks. It stains and it *stings*. The Food and Drug Administration has since banned it use due to the trace amounts of mercury in it, but I still believe in it. This topical antiseptic was 2 percent merbromin solution and 30 percent alcohol, and boy, did it hurt to apply. But I had to get it onto my wounds before bandaging them up for the night. I gritted my teeth and went for it. Tired, sweating, and woozy, but done with my bandages, I collapsed for the night.

The pain was unbearable and if anyone even lightly touched any part of my body, pain shot through it like an electric shock. Tucked in my sleeping bag, I thought to myself, *It can't get any worse*. But, of course, it did.

I woke up early after a miserable night of sleep. We had a schedule to keep and I was determined to go on, but the thought of having to trek back up and over Sawtooth Pass was too much to think about. It took a while to make the decision to move, but I finally stood up—and found I couldn't take a step. My legs ached so deeply inside. Even then, being the determined female that I am, I forced myself to take a step, then another, and then one more. Walking carefully and very slowly, I found that the pain sort dulled a bit. But then came the issue of my pack. Ugh. Now what? Someone had to lift my pack up and hold it balanced in the air as I gingerly slipped the shoulder straps on. As they lowered it onto my back and let go, I thought I was going to die right there and then. *No way. I cannot go on. Impossible!* But as I had done earlier, I slowly took small, gentle steps, and little by little I was able to actually take normal, although very slow, steps. *I can do this*, I told myself. That day's hike seemed to take me forever. I was the very last one to the next night's camping spot, but I made it.

Having overcome both mental and physical anguish, I knew that I would be able to make it through the next twelve days, and I did. I battled off some infection in my shins and I dreaded the morning and nightly bandage changing times, but I pushed through. On day eleven we descended out of the high country, headed back toward the Sierra trails more accessible to hikers who only dared to spend a few nights in the outdoors. I began to see more and more hikers on the trail as I now trudged along pretty well, keeping pace with the others. We were setting up camp for the night in a beautiful, expansive valley. The next day we would reach Mineral King and I would drive home, but we were there for the final night.

It was strange to have other people around us also setting up their overnight spots after being alone—with just our group—for so long. While I was pitching my tent, a U.S. Forest Service ranger came over to greet our group. He was very cordial and had that required, uniformed, "ranger look": brown shorts, brown shirt with a Forest Service patch on the sleeve. After introducing himself, he asked us where we had been, what the conditions were like in terms of snowpack, water, and such; then he glanced down at my now unbandaged legs and exclaimed, "Wow, what happened to *you*?!" I told him briefly about my fall and he replied, "Well, I can radio and get a helicopter in here to evacuate you to a hospital."

"Really?" I responded with pure disbelief. It was all I could do not ask him if he was joking. "It is really not necessary," I said. "Look at my legs, they are healing up *great*." And they were. My wounds were noticeably smaller, although they were stained with Mercurochrome, covered in Neosporin, and did still have just a hint of infection. I couldn't resist adding, "You should have seen my legs ten days ago!"

Two years later, here I was back in Mineral King with friends I had had since elementary school. It was great fun. We made campfires at night and sat around reminiscing about our lives. We *felt* like grownups. It was on this trip during one of my solo day hikes to a nearby ridge top I had an epiphany, one of those transformative moments that I will always remember: *I knew what I wanted to do in life.* At that moment, it didn't point to a specific career choice, as I already had a love for science, but rather *how* I wanted to live my life. I wanted to do something that I loved but also knew that it had to be some*where* I loved. And what do I love? *Nature.* I feel most comfortable with a little breathing room. Whether it be on a ranch, like our family's Mountain House where I kept my beloved horse Vandy, in the Sierra, or at the beach, if I am in a beautiful place in natural surroundings, I am happy . . . no matter the compromises I would certainly have to make to make it happen.

Now, many years later, working in the worlds of science and nature, I am as anxious as I am painfully aware that our natural world is suffering. We are at a crossroads where decisions and actions need to be taken immediately. Many of these imperatives relate to areas of our world that are not so apparent as one might think: under portions of our beautiful oceans are coral reef (colonies of tiny animals) systems. These reefs are the lungs of our ocean and they are dying out at an alarmingly rapid rate due to pollution and warming seas. These coral reef systems are an intricate and a vital part of the health and life of the oceans, which are also part of our atmosphere, and what's happening to them should be "a storm warning" of things to come.

Coal miners used to bring canaries in cages down into the coal mines with them. Somewhere along the line, miners learned that canaries were very sensitive to carbon monoxide and methane. As long as the canaries kept singing, the miners knew everything was okay, but if they stopped singing, this meant that their air supply was no longer safe and the miners needed to evacuate the mine *immediately*. The condition of the coral reefs is analogous to the canaries dying in the caves, except that the coral reefs are more easily ignored. "Out of sight, out of mind." For now. The state of the coral reefs is a ticking time bomb that requires immediate action. We need to focus on them as intently as the miners did with their canaries. At some point, what is happening to our reefs will be writ large for all of us. It is life or death for our beautiful planet.

The atmosphere and the oceans are inextricably linked: both are fluids and both are responsible for the earth's climate and weather. We call it the "ocean-atmosphere" system. The ocean helps regulate the temperature of the lower part of our atmosphere and the atmosphere is in general responsible for the circulation of the oceans through waves and currents. Beneath the surface of these waves, we find coral reefs, which are truly amazing. They are made of extremely ancient tiny invertebrate (lacking a backbone) animals. They range in size from one to three millimeters or a little less than the thickness of your fingernail. But when they grow in colonies, they become huge, and reefs can weigh several tons. Coral reefs are similar in nature to tropical rainforests, due to the diversity of creatures (plants, animals, and fish) that live near and among them. Interestingly, I learned a great deal about coral reefs from an attorney whom I worked with on forensic cases, Murray Camp. He has taken it upon himself to become amazingly well versed in coral reefs and has even published in scientific journals like *Coral* magazine.

These tiny invertebrates evolved into reef builders over twenty-five million years ago! For a long time scientists did not know how to classify these "things." Were they plants, animals, or even rocks? We now know they are tiny animals; there are over six thousand known species of these anthozoans. A coral reef is composed of hundreds of thousands of these tiny creatures. Inside each of these animals is an alga that lives symbiotically inside of it. The coral protects the algae and provides the compounds they need. The coral produces carbon dioxide through its respiration. The coral's waste is composed of nitrates and phosphates. All of these compounds are used by the algae for photosynthesis, using sunlight to create food from carbon dioxide and water. In return the algae produce oxygen and organic compounds such as glycerol, amino acids, and glucose that the coral needs to build and to grow. I find it quite remarkable that there is almost no waste in these solar-powered reef systems. It is the true definition of a symbiotic relationship—you scratch my back and I'll scratch your back, or rather, you scratch my invertebrate and I'll scratch your invertebrate!

If the algal cells are somehow expelled from the coral animal, the coral may die. This expulsion can occur when the reefs are under stress. Coral reef stress can be caused by changes in ocean chemistry, sedimentation, ultraviolet radiation from the sun, and changes in water temperatures. Some of these are natural and some are man-made (anthropogenic). If the algae are expelled from the coral, this is called "bleaching," as the coral now appears completely white. Scientists have discovered that over the past twenty years bleaching of the coral reefs has increased and that global climate change, as well as changes in ocean chemistry from pollution, plays a role.

Coral reefs have a very narrow range of sea temperatures in which they can exist. If there is a drop or a rise in this temperature range, the reefs begin to bleach due to the stress. They are more susceptible to increases in sea temperatures than decreases. An increase of only 2 degrees to 4 degrees F in sea temperature for one to two months can induce bleaching in coral reefs.

Back in 1998 the coral reefs near the Seychelles Islands (off the coast of Africa, north of Madagascar) sustained huge losses due to bleaching, which in turn devastated the fishing in the region. Much of this region's reefs have been unable to recover to this day due to El Niño events that caused the oceans to warm in the region. Another example of how inextricably linked the oceans are to the atmosphere is that the ocean helps regulate the temperature in the lower atmosphere. We now know of the many interactions between the two that contribute to the weather and climate system. Two well-known examples of this interaction are El Niño and La Niña. In a recent interview, Dr. Sylvia Earle, a marine biologist with *National Geographic*, stated that 40 percent of coral has been lost in the past thirty years—coral that took more than 5,000 to 10,000 years to grow. This sudden loss is partly due to the warming of the oceans due to climate change and partly due to carbon emissions, which are altering the chemistry of seawater, making it more acidic. We cannot just cut out carbon emissions—our society as we know it would grind to a halt—but we must significantly reduce these emissions in order to make a serious impact.

"What is your greatest worry about climate change?" Claudia Dreifus of the *New York Times* posed this question to Dr. Jane Lubchenco, marine ecologist and former U.S. Undersecretary of Commerce for oceans and atmosphere. Dr. Lubchenco responded: "I worry about oceans becoming more corrosive, decimating both fisheries and coral reefs. Oceans have already become 30 percent more acidic since the beginning of the Industrial Revolution; if business-as-usual carbon emissions continue, oceans are likely to be 150 percent more acidic by the end of the century. Yikes!" An increase in ocean acidity of this magnitude would not just affect coral reefs, but also impact sea creatures in varying ways. Some creatures, such as photosynthetic algae and sea grass may benefit from the increased carbon dioxide in the oceans, while others see detrimental effects, such as sea urchins, clams and oysters, whose shells dissolve.

The world's most diverse coral reefs are in a "coral triangle" that extends from Bali to the Solomon Islands to the Philippines. Within this region lie Indonesia's Raja Ampat Islands. The Raja Ampat coral reefs make up around 70 percent of the coral reefs globally. They contain more than 1,320 fish species and 540 coral species. There a Marine Protected Area (MPA) has been established. About six thousand of these MPAs have been established around the world. MPAs are protected regions that are set up such that local communities are required to uphold the MPA-established rules. However, the jury is still out on the success of these, as there is conflict between the "local community" and the "outsiders" in using these regions for their livelihood or for profit. Also, it is much easier to protect land regions than ocean areas, as marking territory or tracking animals or fish is difficult in marine regions. This difficulty, combined with the fact that oceans are easy to ignore for most people, is why there are exponentially far fewer protected marine areas than land areas.

The extreme importance of what the coral reefs are telling us about our oceans aside, their beauty cannot be denied. If art and science are linked in the beauty of painting clouds, then there is some equally astounding art right beneath the surface of the sea. Oliver Wendell Holmes, Sr., said it best in his poem, "The Chambered Nautilus"

This is the ship of pearl, which, poets feign,
Sails the unshadowed main,—
The venturous bark that flings
On the sweet summer wind its purpled wings
In gulfs enchanted, where the Siren sings,
And coral reefs lie bare,
Where the cold sea-maids rise to sun their streaming hair.
Its webs of living gauze no more unfurl;
Wrecked is the ship of pearl!
And every chambered cell,
Where its dim dreaming life was wont to dwell,
As the frail tenant shaped his growing shell,

Before thee lies revealed,—
Its irised ceiling rent, its sunless crypt unsealed!
Year after year beheld the silent toil
That spread his lustrous coil;
Still, as the spiral grew,
He left the past year's dwelling for the new,
Stole with soft step its shining archway through,
Built up its idle door,
Stretched in his last-found home, and knew the old no more.
Thanks for the heavenly message brought by thee,
Child of the wandering sea,
Cast from her lap, forlorn!
From thy dead lips a clearer note is born
Than ever Triton blew from wreathèd horn!
While on mine ear it rings,
Through the deep caves of thought I hear a voice that sings:—
Build thee more stately mansions, O my soul,
As the swift seasons roll!
Leave thy low-vaulted past!
Let each new temple, nobler than the last,
Shut thee from heaven with a dome more vast,
Till thou at length art free,
Leaving thine outgrown shell by life's unresting sea!

SEVENTEEN

Extreme Extremes

If you're going through hell, keep going.
—Winston Churchill

Bob had had it. He decided to drop the final t in Dr. John Hallett's name on the class paper he was about to turn in. It now read:

Paper submitted by: Robert Keislar
Submitted to: Dr. John Hallet

It was his only hope. Dr. Hallett, a very well known ice physicist and a research professor at the Desert Research Institute, was teaching a course in atmospheric instruments and kept spelling Bob's last name as Keisslar, with an extra s. Though Bob had mentioned this numerous times to Dr. Hallett, it never sunk in.

In the very next class, Dr. Hallett mentioned the misspelling in his own name right away, speaking directly to Bob in front of the entire class. "You spelled my name incorrectly!"

Bob responded very politely, "When you take the extra s out of my name, I will add the t back to your name." And that was all it took. Neither misspelled the other's name again.

This story illustrates an important issue that scientists face in getting people to understand the gravity of the situation that our Earth is now facing. Scientists cry out for more awareness and action regarding climate change, pollution, and other problems, yet their pleas often fall on deaf ears until a particular group is directly affected in a negative way by one of these issues. Extreme weather events have been occurring with seemingly greater regularity than in the past, and this, combined with the Internet spreading almost instant notification globally, seems to have gradually caused people to pause and think about what is occurring on our planet. This awareness must increase in order for us to face this global challenge with the force and action we need in order to start realistically dealing with the effects of climate change. Why? Because we don't have a choice if we want to keep the Earth alive and well. Hurricanes, like Sandy and Katrina, devastating tornado outbreaks, droughts, floods . . . extreme weather events may not be occurring more frequently due to climate change, but they will *be* getting more extreme.

Climate change is not limited to the fact that the global surface air temperatures are warming or that the stratosphere is actually cooling. Climate change has been blamed for the fact that extreme weather events are increasing in intensity and perhaps in frequency. To whatever cause one attributes the increasing intensity of extreme weather events, many scientists, including me, now use the term "the new normal" to describe this scenario. Around the world this "new normal" is finally being realized. The United Kingdom's Meteorological Office, their national weather service, has documented an increase in extreme events in the U.K. One area in particular is rainfall. A half-century ago, rainfall used to reach the top 1 percent of the average amount for that time of year every one hundred days. Now this happens every seventy days. Some places around the globe are seeing colder temperatures and more snow, while others are experiencing more intense heat waves and droughts—all extreme events. Since 1950, every decade in Australia has been hotter than the one before it, as noted by scientists at the Australian Commonwealth Scientific and Industrial Research Organization (CSIRO). Much of Queensland, Australia, and portions of southwestern New South Wales are still experiencing record drought. Jerusalem saw its biggest snowstorm in twenty years in 2013 when eight inches of snow fell on the city. The list goes on and on.

The biggest worldwide weather killer is heat. It may not have the drama of a hurricane, but heat waves actually kill more people than floods, lightning, tornadoes, and hurricanes combined! In the United States, the death toll from heat is followed very closely by cold weather deaths, with an average of 660 deaths per year in the U.S. due to heat and in some years cold kills more than heat, however, overall, the statistics show that heat edges out cold in the U.S. In July of 1995 the city of Chicago was impacted by a terrible heat wave in which seven hundred deaths were blamed on the heat.

The top six weather killers in the United States, based on statistics from various federal agencies, are:

1. Hot weather
2. Cold weather
3. Floods
4. Tornadoes
5. Lightning
6. Hurricanes

Though heat and cold deaths are nearly tied in the U.S., as mentioned earlier, Third World countries see much higher death rates due to floods and windstorms. This is attributed to inadequate preparedness for and warnings of these weather events. Many of these deaths in less developed countries are indeed from hurricanes. If these storms occur in the South Pacific or Indian Ocean, they are called cyclones. If they occur in the northwest Pacific they are called typhoons. But they are just different terms for the same thing. A hurricane is a tropical cyclone with sustained surface winds of at least 64 knots (approximately 74 miles per hour) that occurs in the Western Hemisphere (Caribbean Sea, Gulf of Mexico, North Atlantic

Ocean, and in the eastern and central North Pacific—east of the dateline). The name *hurricane* comes from the Mayan storm god name "hurrican," a Taíno word. The Taíno language was spoken by the Taíno indigenous peoples of the Caribbean; there is dispute about whether or not this language is officially extinct. The first official use of the term *typhoon* was most likely when it was published back in 1560 in the Portuguese officer Major Serpa Pinto's journal, where it appears as the Portuguese word *tufaõ* and Pinto claims that this is a word used by the Chinese to describe these types of storms. But there is debate as to the original term, as it could also have been derived from the Cantonese word *t'ai fung*, or a "great wind," or from the Greek word *typhon*, which means "monster." I tend to side with the Greek origin of the word, as Aristotle used this term in his book *Meteorologica III*, which he wrote in 350 BC, to mean "wind-containing cloud."

Regarding these "monsters," whatever term is used, I will use the term *hurricane* to represent all of them going forward. In the northern hemisphere, hurricanes rotate counterclockwise, and they rotate clockwise in the southern hemisphere due to the rotation of the earth and the "apparent force" called the Coriolis force, which pulls winds to the right in the northern hemisphere and to the left in the southern hemisphere. These hurricanes are fueled by deep layers of warm waters and obtain much of their energy from the latent heat that builds up inside them and is then released due to condensation from the numerous large thunderstorms contained within them. Latent heat is energy absorbed or released by a substance during a change in its physical state while its temperature remains constant. Hurricanes differ from mid-latitude cyclones in that mid-latitude cyclones are larger in size than hurricanes and these cyclonic storms have a cold core. Hurricanes like Katrina have warm cores and are typically 350 miles in diameter, whereas mid-latitude cyclones can be 1,000 to more than 3,000 miles in size. The center of a hurricane's pressure can be 990 millibars (or hecta Pascals) but is typically 950 millibars, although in the eye of a typhoon off Guam on October 12, 1979, the pressure dropped to 870 millibars (or 25.69 inches of mercury), the lowest on record.

Hurricanes are the most powerful of all storms, although tornadoes have the strongest wind speeds. Hurricanes are described by their "category" and measured on the Saffir-Simpson Hurricane Wind Scale, which ranks them from the weakest—category 1, with sustained winds of 74 to 95 miles per hour, where some damage will occur—to the strongest—category 5, with sustained winds of 157 miles per hour or higher, when catastrophic damage will occur. Category 3, 4, and 5 hurricanes are considered to be *major* hurricanes.

There are many factors that play a role in forming a hurricane. Dr. Christopher Landsea, the Science and Operations Officer at the U.S. National Hurricane Center, and I attended UCLA's undergraduate atmospheric sciences program together. Though most articles one reads these days claims that hurricanes will increase in strength and frequency due to "global warming," Dr. Landsea makes an excellent argument for the case that, though hurricanes may be slightly stronger (by a few percent), the number of tropical storms and hurricanes will not go up, and may even decrease. Chris is not a climate change disbeliever, but he understands the components that go into making a hurricane and how these are affected by warming oceans and atmospheric temperatures. This explanation makes perfect sense to me.

Dr. Landsea lists the "ingredients" for hurricane creation as follows: moist air, numerous thunderstorms, weak vertical wind shear (between surface and upper level winds), and some sort of triggering mechanism, such as boundaries between masses of warm and cold air disappearing or a convergence of the northern and southern hemisphere easterly trade winds. Chris explains his conclusions:

> *It's also important to point out that ocean temperatures are not the only factor that is crucial in knowing which disturbances will develop into a tropical storm and which systems will intensify to become extremely strong hurricanes . . . Any manmade alterations to the air's moisture, thunderstorm activity, vertical shear, and originating disturbances may be as or even more important than changes to the ocean temperatures themselves. All climate models predict that for every degree of warming at the ocean that the air temperature aloft will warm around twice as much. This is important because if global warming only affected the earth's surface, then there would be much more energy available for hurricanes to tap*

> *into. But, instead, warming the upper atmosphere more than the surface along with some additional moisture near the ocean means that the energy available for hurricanes to access increases by just a slight amount. Moreover, the vertical wind shear is also supposed to increase, making it more difficult (not easier) for hurricanes to form and intensify.*

Though hurricanes' intensity may only increase by a few percentage points and their frequency will remain constant (or may even decrease), there is ample evidence that other forms of extreme weather events are increasing in intensity *and* frequency. This also illustrates the complexity of the climate change science and why it is important for scientists—and not politicians or biased industries—to be in charge of research programs and debate over these issues as they continue to develop over the decades to come.

The headlines today *should* read: THE NEW ABNORMAL COMES TO HOLLYWOOD: WEATHER EXTREMES AND CLIMATE CHANGES IMPACTS THE STARS. Maybe then people would listen. *No one* is immune to these changes in weather patterns, no matter how rich or poor. The growing frequency and intensity of extreme weather conditions around the globe has begun to threaten the glamorous and exclusive multimillion-dollar beachfront properties of Hollywood's rich and famous as their beaches are being eroded at an increasingly rapid rate. Many stars have and have had homes on Broad Beach Road, an area that's been affected, including Robert Redford, Dustin Hoffman, Mel Gibson, Robert De Niro, and Steven Spielberg, to name just a handful.

When my parents moved our family—me, my sister, Greer, and my brother, Patrick—from New York to southern California in 1969, we were small children all under the age of five. My parents rented a house on Broad Beach Road right on the ocean in Malibu. Malibu is quite long and extends for a twenty-one-mile strip along the Pacific coastline in Los Angeles County. It begins to the south at the town of Pacific Palisades and goes up to the Los Angeles/Ventura County line. My parents ended up buying a home in Santa Monica to the south of Malibu, as back then Broad Beach was out in the boonies. Also, there was no high school there at the time, although I attended Malibu Montessori School while we lived out there in the sticks. Broad Beach Road lies toward the very north end of Malibu, on the north end of a beach called Zuma Beach. Zuma Beach is notorious for large waves and great surfing. The reason it gets much larger and stronger waves than other locations in Malibu is that it faces the open ocean to the southwest. Most of the other beaches in Malibu lie in coves that tend to protect them, and their beaches face the southeast. What this means for the owners of beachfront homes on Broad Beach Road is that they are exposed to the storms, wind, and waves that impact southern California. Thus, they get much more beach erosion than the other beaches in Malibu.

According to recent reports, a group of Broad Beach homeowners has banded together to pay for a potentially costly project to bring sand in to help alleviate the beach erosion. Sand is being brought in from inland sources such as the Mojave Desert and offshore sources such as the sand off Dockweiler Beach (which is near the Los Angeles International Airport).

The fact that Malibu's Broad Beach is more exposed to the open ocean will make the success of the sand replenishing efforts more difficult than on other beaches, as Broad Beach is more susceptible to sea level rises, along with the possible impact of more frequent extreme weather events. As the sea level rises (which it will), the damage that can occur from weather events (not even extreme ones) will become more severe.

There is a history of both successful and unsuccessful sand-replenishing efforts around the U.S., much of which occurs in the state of Florida, where almost 60 percent of state beaches are experiencing beach erosion. Sand replacement is very expensive and environmentally questionable. Florida spends approximately a hundred million dollars annually on sand replenishment projects and, regrettably, the beach will most likely erode again in a few years and the process will have to be repeated.

But the loss of beaches is just one small facet of the "new normal." Now that we know the surface temperatures of the Earth are warming and that climate is changing, there is no doubt that many extreme weather events will continue around the globe—with greater severity and, in some cases, frequency. These include

- Flooding and snowfall occur in the Sahara Desert.
- China just experienced its coldest winters in nearly thirty years.

- Eastern Russia is reporting temperatures down to –50 degrees F, which is a record low.
- Heat waves and brush fires are occurring in parts of Australia that have never experienced them before.
- Rio de Janeiro reported a temperature of 109.8 degrees F on December 26, 2012, the highest temperature reported since record-keeping began in 1915.
- Heavy snowfall occurred in Tokyo, which usually only experiences light dustings.

If the recent extreme weather conditions in Southern California, like flooding and fires, are a snapshot, Hurricane Sandy was an urgent wake-up call to the U.S., and there will be another "Hurricane Sandy" that will threaten us again. For a hurricane to reach as far north as Sandy did was an unprecedented event. Even if hurricanes, as Dr. Landsea predicts, will not increase in frequency, the damage to our coastlines is more serious than people realize. This damage can occur not just from hurricanes but also from tsunamis (recall the lives lost in the 2004 tsunami caused by an earthquake in the Indian Ocean, killing an estimated 230,000 to 280,000 people), large storm events, and rising sea levels.

Scientists' cries for more awareness and action regarding climate change, pollution, and other problems often fall on deaf ears until a disaster strikes. I am keenly aware of extreme weather and its impact and effect on people, planet, and the environment; this is why I named my company WeatherExtreme Ltd. Though climate change may not cause extreme weather events, it does cause *more* extreme weather events to occur. This is a fact! More and more beaches will begin to erode and the ones already eroding will erode more quickly. There will be more tornado outbreaks, more droughts, more floods. We discuss hurricanes so much due to their impact on such a grand scale and the incredible destruction they cause for years after the event.

Why are hurricanes hitting places they never did before? Why are we so vulnerable? What problems with city infrastructures are exposed? What about the massive erosion on our shores, the destruction of cities, and the deaths that keep occurring even though our weather forecasting has improved tremendously? One interesting fact that *is* encouraging: forty years ago, the average three-day forecast of hurricane landfall was off by four hundred miles; today our average forecast is almost down to just an eighty-mile margin of error! Nevertheless, the paradox remains that we are living in an increasingly wealthy country and yet our cities' infrastructures are old and dilapidated, so that when they are impacted by extreme weather, the destruction runs into the millions of dollars very quickly, despite the great advances in warning. These issues need to be addressed by the general public with the guidance of concrete science—concrete science, unfortunately, resides in a world that is polluted with so much "junk science" that one often requires a guide to help make the distinction between the two. (I have to emphasize the word "concrete" as opposed to "junk" science, the latter being riddled with errors and biases to suit a commercial or political interest.) Otherwise it is like shoveling sand against the tide . . . just ask the beachfront home owners in Malibu's Broad Beach in ten years' time.

EIGHTEEN

Forensics, Extreme Wx, and Death

The world is full of obvious things, which nobody by any chance ever observes.

—Sherlock Holmes,
The Hound of the Baskervilles,
SIR ARTHUR CONAN DOYLE

Feeling uneasy due to the crime statistics of the area, I pulled to the side of the road by the house where the brutal murders took place. My nondescript Hertz rental car didn't stand out, but I did as soon as I stepped out of the car. I immediately felt eyes staring at me, although I didn't immediately see anyone. I could hear what they were saying: "What's that white lady doing poking around the murder scene?" I thought that a little strange, seeing how it was long after the murders took place, but the crime clearly was still resonating within the community. Luckily, shortly after I got out of my car, the attorney who hired me pulled up and parked just behind me.

The Mead Valley could be described as a lower economic area of Riverside County, California. It is located in what is ironically called the "Inland Empire." As far as I could see, the only accurate part of that description was that it was inland. As I drove down the streets I noticed that some of the homes in this Inland Empire community were well cared for, but others had junk stored all around them and plots of weeds posing as lawns. There was a mixture of stick-built houses, manufactured homes, and trailers. The house where the murders took place was a rather generic-looking trailer but there was nothing generic about what occurred inside that trailer. It was the very early morning hours of December of 1993 as a man and a teen entered the home of an elderly couple, intending to rob them. But it went horribly wrong. Apparently, the couple "didn't cooperate" so the robbers hog-tied them with wire, leaving the tip of a latex glove in one of the knots on the wire. They stabbed the couple multiple times and left them to die.

It was late afternoon the day I came. It felt eerie walking around the crime scene and the nearby locations I needed to see, so I got to work to try to take my mind off the tragedy. I took measurements of the height of the surrounding terrain with my inclinometer, and various readings with my compass. There was a noticeable lack of streetlights—a perfect environment for criminals. An eyewitness to the crime saw two people entering the yard (one going in the front yard and one going in the backyard) of the murder victims' house. The eyewitness was a young boy who was up very early and noticed the defendant's car drive by the elderly couple's house. The boy was across the street in a travel trailer–type vehicle. So I needed to determine the physical conditions: the surroundings, the elevation, the terrain, and the orientation of the house compared to where the eyewitness was located.

In this double murder case for which the death penalty could potentially be sought, visibility was one of the key issues. I had to determine the amount and nature of the natural lighting conditions on the day and at the approximate time of the murders. In order to do this, I needed to know as much as I could about the location, the weather on the day and approximate time in question, as well as the sun and moon data. So a site visit was extremely important. Cloud cover, air temperature, relative humidity, and weather conditions also play a huge factor in determining visibility. During the time of the murders, there were just a few high scattered clouds, there was no moon in the sky, and the sun was below the horizon.

During the early morning and late evening hours of every day on Earth there occur three types of twilight. In the early morning hours the first twilight to occur is astronomical twilight, then nautical twilight, and then civil twilight. Twilights are not an instant in time but rather occur over a span of time of approximately thirty minutes. During this time period, the sun is slowly making its way closer to the horizon for sunrise so that during each twilight period of time, the natural lighting conditions are changing.

The actual definition of each twilight time is defined by the number of degrees the sun is below the horizon measuring from the geometric center of the sun. Astronomical twilight begins in the early morning hours when the sun is 18 degrees below a flat horizon; nautical twilight begins when the sun is 12 degrees below the horizon, and civil twilight begins when the sun is 6 degrees below the horizon. Sunrise then occurs when the center of the sun is on the horizon. For evening twilights, these figures are just reversed to define the ending of these twilights: that is, civil twilight ends when the sun is 6 degrees below the horizon; nautical twilight ends when it is 12 degrees below the horizon; and astronomical ends when it is 18 degrees below the horizon.

All of the twilights, as well as sunrise and sunset, are defined with a flat horizon. This is to keep things consistent. But in many places around the globe the horizon is not flat. This is where a site visit to a specific location becomes very important.

During civil twilight, either in the early morning or the early evening, it is still possible to work out of doors without artificial light for occupations such as road work. During nautical twilight, the human eye begins to see things only in black and white, not color. "Nautical twilight" earns its name as one can still sail ships and make out large rock formations for navigation. Since each twilight lasts about thirty minutes, obviously the natural lighting conditions are changing, either getting lighter or darker, depending on whether it is the morning or evening, at a fairly fast rate.

As I reviewed the documents regarding the grisly murders, it appeared that they occurred during astronomical twilight. During this twilight it is very, very dark without any artificial lighting. The stars are brighter than the skylight. From the site visit of the murder scene, I made accurate calculations of the various beginning and ending times of the twilights, given the terrain surrounding the location. I also gathered all of the weather data from surface weather stations, upper air balloons launched into the atmosphere, and satellite and weather radar data. I analyzed all of the data and information to come up with my opinions in the case, including details about the times of the various twilights (position of the sun below the horizon), the cloud cover status, and any temperature inversions as visibility may be greatly reduced due to accumulation of dust and pollution even without any clouds being present.

A few months after my site visit, I sat on a bench outside the courtroom with my large notebook containing my complete case file on my lap. The bailiff opened the courtroom door and let me know that it was my time to testify. As I entered the room, I scanned the overall scene. I already knew who the jurors were in terms of their careers and who was the foreman. I didn't yet check out the judge or the defendant.

"Please approach the witness stand to be sworn in," said the judge. "Reporter, please swear in the witness."

The court reporter swore me in and the judge asked that I take the stand. While taking my seat in the witness stand, I organized my file in front of me and prepared for one of the two defense attorneys to begin my direct examination. Then I scanned the courtroom again, now from the perspective of the stand. It was very large courtroom and yet not very full. I turned my head and glanced at the defendant, who was the older of the two arrested for the murders. To date, I had never laid eyes on him. He was a well-built African American man sitting behind the defense table.

"Please state your full name for the court," the defense attorney said, and I turned to him and got to business.

The defense attorney asked me questions about the weather, sky cover, and the natural lighting conditions that were present during the early morning hours of the day in question. I described the twilights and their definitions in detail to the jury, making sure that many of them were following my explanations. If a few of them understood, they could always explain it to the rest, if necessary. I asked the judge if I might write on the whiteboard that was between the jury and me. He agreed.

I meticulously wrote down each of the twilight times on the whiteboard for the jury to see. Every time I answered a question from the attorney, I turned to the jury to answer that these are the "trier of facts." After cross-examination by the prosecution and no further questions from the defense, the judge excused me.

I found out later that the timing of the murders was the subject of great debate during the trial. The entire debate was over a one-hour time period during which the murders possibly occurred. This one-hour period spanned from astronomical twilight halfway through nautical twilight! This meant a big difference in the natural light available for the human eye to see, thus potentially calling into question the testimony of the eyewitness. I also learned that there was testimony about a security light that was apparently near the house. Either way, I testified to the science of the conditions during the early morning hours of these brutal events, and then the jury would use the information as they saw fit.

The jury returned the verdict of guilty, with the death penalty.

Through the years I have worked a wide variety of cases, but this double murder, death penalty case stood out due to the enormous consequences. One of the experts I worked with on several cases over the years was a man named Richard "Pete" Burgess, a Texan through and through, in the best sense of the word. Pete is an aviation consultant with around forty years of experience as an air traffic controller, supervisor, and branch manager in the Navy and the Federal Aviation Administration.

Early on in my career, Pete said to me, "Always remember, experts don't lose cases; attorneys lose cases."

To this day I often think of his advice. I also realize that the opposite is true: experts don't win cases; attorneys do. And it is our job as expert witnesses to provide the facts without agenda so that the attorneys can present their respective arguments accordingly. It is hard not to become emotionally invested, sometimes, but I have to remind myself to heed Pete's advice and check my own personal feelings at the courtroom door.

The skies were darkening and the winds were picking up on a Florida beach one hot August day. Two teenage sisters walked out to a boat that was waiting to take them on their parasailing adventure. The girls had been watching people parasail all day and were so excited to try it. They were vacationing with their family but their mother did not come along to watch this particular adventure. They called their mother, who was back home, and begged her to let them go parasailing. She finally agreed.

As they strapped into their parasailing harnesses, little did they realize that a just over a half hour earlier, the Miami National Weather Service Office had issued a Marine Weather Statement for the area, stating that there was a line of thunderstorms detected on Doppler radar capable of producing winds of up to 30 knots (about 35 miles per hour). In addition, they stated, "As thunderstorms move over the water . . . boaters can expect strong, gusty winds . . . high waves . . . dangerous lightning . . . and heavy rain."

The photographs and video I analyzed as part of my meteorological investigation of the accident showed that as the two girls walked to the boat that was preparing to launch them to the skies, the skies in the east and southeast were rapidly getting darker. This was the line of thunderstorms moving in on the area that the National Weather Service had just warned about.

During the girls' parasailing flight, the winds picked up significantly and they became frightened. One of them yelled down at the captain to pull them in. But the ride continued.

The boat captain told the sheriffs who arrived on scene after the accident that the winds had suddenly increased from 15 miles per hour to approximately 40 miles per hour. As a result of the increasing wind, the hydraulic winch on the boat was no longer able to pull the girls down. This is most likely why, unbeknownst to the girls, their ride continued even after they started shouting to come down. Not only did the now-unwanted ride continue, but the winds were pulling the boat toward shore, until the boat became grounded on the beach and the girls hung precariously over the beach, not the water. This lasted for about two minutes. The rope to the parasail then snapped and the girls flew out of control, west toward the row of hotels and villas that lined the beach.

There was nothing onlookers on the ground could do but watch in horror as the parasail with the girls spun out of control. The winds blew the parasail with the girls hanging from it into the rooftop of a building and dragged them across it. The wind then took them through several trees, leaving them suspended in the air, hanging over a courtyard, both still alive, one barely clinging to life, with a broken neck.

Meteorologically, this was a typical Florida summer afternoon, where showers and thunderstorms develop during the heat of the day, moving through parts of the state very rapidly. There was an upper level area of low pressure moving across the region providing the atmospheric support necessary for the development of thunderstorms. After noon that day, the air temperature dropped almost 8 degrees Fahrenheit in about an hour and a half, though it was still the middle of the afternoon. The wind speed then picked up as the thunderstorms approached the region of the parasailing activities. These thunderstorms impacted about a sixty-mile-long stretch of the Florida coast, including the parasailing location. With them, they brought cloudy skies and gusty winds. In addition to easily obtainable weather forecasts, any basic satellite or NEXRAD radar image would have depicted the area of thunderstorms moving in on the region from the southeast over the Atlantic Ocean.

There are a number of ways to query weather data and reports: via the radio, television, NOAA Weather Radio, Internet, and telephone. During the time of this accident, there were no laws (federal or state) governing parasailing operations in Florida. There were just "industry guidelines" stating that it is the captain's responsibility to evaluate and determine if weather conditions are favorable for parasailing and that "the Captain shall use all means to make such a determination." One of the guidelines was "that no vessel shall be operated if the winds were in excess of 20 mph or when excessive or dangerous wind gusts are present." The captain, doubtless in his own defense, stated that the winds picked up speed quite suddenly. One would logically assume that any experienced captain would be quite familiar with the dynamic nature of weather—and certainly the very dynamic nature of Florida weather. Analyzing the case as a forensic meteorologist, it was clear to me that weather did not "sneak up" on them. Though the morning weather forecasts called for winds of only 5 to 10 knots (about 6 to 12 miles per hour) long prior to issuing the marine statement about approaching thunderstorms, the early morning (just after six A.M. local) National Weather Service area forecast stated:

> ... STRONG THUNDERSTORMS POSSIBLE FOR SOUTH FLORIDA TODAY ...
> ... WIND GUSTS TO 40 MPH AND DIME SIZE HAIL POSSIBLE WITH STRONGEST STORMS ...
> THIS HAZARDOUS WEATHER OUTLOOK IS FOR SOUTH FLORIDA.

Meteorologically speaking, this forecast certainly should have put one on alert. In addition, NOAA weather radio not only broadcasts the forecasts but any marine weather statements and changes in the weather. No matter what one is doing, especially activities outdoors, it is important to remember, "Weather is dynamic"—sometimes dangerously so!

A recent National Transportation Safety Board report stated that between three and five million people a year participate in parasailing activities in the United States alone. About one-third of this activity takes place in the state of Florida. Ninety-five percent of parasailing fatalities result from the parasailors' inability to escape from the harness following an unplanned water landing in high winds.

To date, there are still no federal regulations or guidelines for certification or specific training of parasail operators, which, given the extreme risk inherently associated with that activity, was extremely surprising to me when I first learned that fact. Nor is there any federal requirement for inspection of parasail equipment nor any requirement to suspend operations during inclement or unsuitable weather conditions. The NTSB has recommended creating a special license class for commercial parasail operators, but as of the date of this book's publication, nothing has yet come to fruition.

Regarding state laws, only the states of New Jersey, Virginia, and as of October of 2014, Florida, have some form of minimal standards for parasailing equipment and operational limits due to wind and sea conditions.

The Florida Parasailing Act, also known as the White-Miskell Act, was created and named after the fifteen-year-old killed in the accident I investigated, Amber White, and for a tourist, Kathleen Miskell, killed in Florida when her parasailing harness broke and she fell into the water. This new Florida law is quite stringent, especially in terms of weather, as it prohibits commercial parasailing during any of the following: sustained winds of 20 miles per hour or higher, wind gusts greater than 25 miles per hour, rain or fog diminishing visibility below a half mile, or lightning storms detected within seven miles. I often wonder, if the girls' mother had been there at the beach when the girls were begging to go parasailing, would she have allowed them to go once she had seen the darkening skies looming?

Sometimes I am hired to investigate the weather in order to decide it if played any role at all in an accident or death. Sometimes it does not, but that too has to be explained to the jury or the judge, in order to rule it out completely and allow them to focus on other causes, whether human error or something more malicious, as with the aforementioned instance of pilot suicide. Other times, however, weather does indeed play a crucial role in a case.

"I tripped and fell on some boards because I couldn't see them, as they were all covered up with all the new snow that had just fallen," explained a propane company worker. The investigator for the propane company's insurance company called me to research the weather on the day this worker reportedly fell and hurt himself on the job. The man was filling a propane tank in the backyard of a house in Truckee, California. Truckee lies about ten miles north of Lake Tahoe. It is notorious for its cold temperatures during the winter months and does receive quite a bit of snow, about 200 inches per year. This is a lot of snow when you compare it to the U.S. average of only 25 inches per year. So the propane worker probably felt quite safe in the claim of all the new snow.

However, things proved not to be so simple. "Not only was there no new snow on the date of this fall, but there hadn't been any new snow in Truckee for almost two weeks," I explained to the investigator. And with that news, the insurance company promptly declined the claim.

Most forensic weather cases are not as clear-cut. Automobile pileups where fog is involved can be quite complicated. These cases also depend on the laws of the states in which the accidents occur, along with the circumstances surrounding the accidents. Was the fog forecast? Should the state's Department of Transportation have been aware of the possibility of fog and put signs out warning drivers? And the list goes on.

Many places around the world have names for their "local" fog. In California one of the big culprits is called "Tule fog." This fog is notorious for forming during late fall through early spring, when high pressure dominates the San Joaquin and Sacramento Valley regions. At times this radiation-type fog (produced over land due to radiation of heat from the land out to the air usually beginning at twilight and occurring during the nighttime hours under light to calm surface wind conditions) can last for weeks. It can be very depressing. I lived in the Davis, California, region for a year and at times I didn't see the sun for a couple of weeks due to this fog. Growing up in mostly sunny Santa Monica, I didn't realize how much the weather affected me in so many ways. When the Tule fog would last more than a few days, I found that I didn't feel like going outside and exercising, I didn't care as much about the types of foods I ate, and I became a bit depressed. I was never depressed before this time. It dawned on me how powerful the weather is in affecting my entire well-being as well as the well-being and attitudes of many people around the globe. Seasonal Affective Disorder (SAD) in places where sunlight becomes very scarce (or nonexistent) for long periods of time is a very real ailment.

It has been proven that weather is associated with outbreaks of influenza, pneumonia, bronchitis, and other illnesses—during increases in temperature we see an uptick in these diseases. When it is very stormy outside, there are increases in birth rates; hot weather can cause decreases in sperm counts, and cold weather can affect morbidity and mortality rates. Gloomy weather, as nonspecific as that term might sound, can have real effects on mood, too, but, interestingly enough, however, suicide rates are not the highest during the coldest, darkest winter (gloomy) months but during the spring months. The consensus is that this is because during the deepest "depression" times of the winter months, people are too depressed even to attempt suicide. But when spring rolls around, the depression does not necessarily "end," but changes, and this can be a catalyst for suicide.

Coming back to fog, I have worked too many roadway automobile pile-ups involving fog to keep track. Usually they are extremely horrific, due to the nature of a vehicle traveling at highway speeds crashing into cars going very slowly or completely stopped. As I inspect the photographs from the accident scenes for clues to the weather conditions that existed, especially the visibility, wet roads, drizzle or black ice, I come upon bodies halfway out of their vehicles, portions of bodies in the roadway, and people left to perish underneath tractor-trailer rigs. It is awful, but I never know which images will provide the weather information I need, so I analyze all of them. I reflect on why I like to drive larger vehicles, as they are more visible to others on the road and safer in an accident. Many people falsely claim that Sir Isaac Newton's Third Law of Motion—when a rigid body in motion exerts a force on a second rigid body, the second body exerts a force equal in magnitude and opposite in direction to the first body—means that small cars are equally as safe as large ones. This is simply untrue. I know it doesn't take into account airbags, crumple zones, and such on a vehicle, but consider a large insect, like a locust and a semi truck. The locust hits the windshield of a semi truck driving down the road. The insect impacts the windshield with a certain amount of force and the truck exerts an equal but opposite force on the locust. The bug winds up squashed all over the glass yet the truck is unharmed, other than needing its windshield washed. Yes, disregarding a few assumptions, the forces were equal and opposite but the semi truck, with its greater mass, was much more capable of handling these forces than the insect. And so this is how I feel when I drive my sport utility vehicle. It is a trade-off I consciously make for driving a vehicle that uses more gas and thus contributes more to the environmental problems we face versus my safety and the safety of my family; there are too many "nuts" on the road. Oftentimes, my ultimate conclusions on traffic accident cases are that the weather was not the primary factor in a crash. But careless driving and not paying attention to changing conditions usually are the likely causes. Luckily, now there are sport utility vehicles that are hybrids.

Some of the vehicle pile-ups in the California central valleys during Tule fog events are mind-boggling in their scale. In February 2002, there was a *ninety*-car pile-up with two killed. I was hired to calculate the weather and visibility conditions on the roadway. I take into account the orientation of the roadway, and the position of the sun, along with nearby airport visibility and cloud cover measurements, to make my calculations. Eyewitness accounts tell of people driving much too fast for the conditions. Others say the fog came up all of a sudden out of nowhere. This is the nature of eyewitness testimony. I've found that it must be taken with a grain of salt, so to speak.

There are many types of fog, including radiation fog that forms on a clear, cool, calm night, like the Tule fog. There is advection fog, upslope fog, sea fog, ice fog . . . and the list goes on. There are many more nicknames for local fogs that occur all around the globe. When it is foggy, I enjoy the quietness of it outside. It seems to dampen all noises. It reminds me of one of my favorite poems, called "Fog," by Carl Sandburg.

The fog comes
on little cat feet.

It sits looking
over the harbor and city
on silent haunches
and then moves on.

But fog can be deadly, make no mistake about it. Fog from pollution can cause death. Fog and pollution together is now called smog. The term was first used back in 1884! Smog was coined to mean a city's fog that is made darker and heavier by smoke and atmospheric pollutants. Hence the combination of "smoke" and "fog" to make "smog." In the 1920s, London protestors tried to get the government to regulate the burning of coal, but these cries fell on deaf ears. Then, in 1952, there was the case of the notorious London fog. Fog in London is quite common, but in November and December of 1952 it turned quite deadly. The weather was colder than normal, so people began burning fires in their chimneys, the majority using coal, as did industry in the region. Industry spewed sulfur dioxide, as well as nitrogen oxides, mercury compounds, and carbon dioxide into the atmosphere. Sulfates and nitrates cause significant health problems to humans and animals. (China is currently the world's largest consumer of coal and is feeling its effects with over 30 percent of the county experiencing acid deposition as a result of these sulfates and nitrates being pumped in to the atmosphere.)

On December 5, 1952, a deadly combination of events occurred over London. An anticyclone, or high-pressure area, moved in over the region. This trapped the cold air over London along with all of the smoke and particulates from the burning in the chimneys. The smog had been bad since November, but now it was exacerbated and clogged the air. A couple of days later the visibility was down to just about one foot. In additional to the particulates in the air that caused breathing problems, the pollutants from burning the coal interacted with the fog, that is, the cloud water droplets, forming sulfuric and hydrochloric acid, which caused horrible burning of people's eyes and lungs.

This horrible smog also caused a crime spree in London. Since it was so difficult to see, people were attacked and robbed on the streets. Cars were abandoned all over the city as no one could see where they were driving. Hospitals were full of people struggling to breathe and dying painful, frightened deaths. The air was poisoning them all, and there was no seeming escape.

It all came to an end quickly as on December ninth, the wind picked up and the fog dissipated. But the damage was already inflicted. The final death toll came in at an estimate of more than 12,000 souls!

During the 1970s, a teacher of mine described a recent visit to the doctor when she was not feeling well. She did not have a fever or cold, but just a general malaise. She would go from feeling fine, to not so fine, on and off, for months. The doctor examined her and could not find anything wrong with her. Luckily, the doctor asked all sorts of questions about her daily routine, diet, and exercise habits.

"I eat well, get eight hours' rest every night, and exercise on a regular basis," she answered.

"Do you fall asleep with the television on at night?" the doctor asked.

"No, I generally read for about an hour, then turn off the lights and fall asleep," she replied.

"What do you do to exercise—for how long and how many days a week?" the doctor continued his questioning.

"I swim laps about once a week at the school pool and I jog around my neighborhood around two times a week," she continued, "and I play golf two or three times a month on weekends."

"Tell me more about your running. What time of day and for how long do you run?" he asked.

"Due to my teaching schedule, I usually run in the evenings. Also because it is cooler in the evening," she replied. She lived in the San Fernando Valley of southern California, which can get quite hot during the days, especially during the summer and fall seasons.

The doctor was quick and emphatic to respond, "I want you to stop jogging at night, and if you do want to run, only run during the day!"

"What? Why?" This would throw her entire routine out of whack.

"Because once it is dark, the industries in the region all spew out the pollutants they don't want people seeing or smelling during the day. You are not the first patient of mine that has come in with complaints of not feeling well but testing fine when I examine you," he explained.

She immediately stopped all jogging in the evening hours, and all of her ailments and general feelings of not feeling well vanished.

The San Fernando Valley is like many valley basins, where on calm nights, the cold air from the surrounding region sinks down into the basin. This traps the air, and whatever is in it, inside the valley. If pollutants are released into the atmosphere during this time period, they subsequently cannot escape the basin, which creates toxic conditions. This was exactly what was occurring back in the 1970s, when my teacher kept falling ill, not just in the San Fernando Valley, but in many other locations around the United States. Today, thankfully, there is now monitoring of air pollution 24/7 in most cities across the U.S., and the Environmental Protection Agency (EPA) will track down and fine these industries for their illegal practices.

Weather can also affect forensic science and postmortem investigations. There are three main stages a corpse goes through after death: primary flaccidity, rigor mortis, and secondary flaccidity. Usually during the first one to two hours after death (although it can last up to six hours, depending on the condition of the body and the environment) primary flaccidity occurs. That is, all of the muscles in the body relax and the body flattens out. Also within thirty minutes after death, livor mortis, or hyperstasis, begins: after the heart stops beating, blood begins to pool in lower portions of the body due to gravity. After about another half hour, staining or purple areas appear on the portions of the body that are closest to the earth as the blood settles. This is called lividity. Lividity is an important tool in determining the time of death.

The second phase begins two to six hours after death and is called rigor mortis. This refers to the hardening of the muscles due to the lack of respiration, depleting the body of oxygen and allowing rigor (stiffening) to begin. Rigor begins in all muscles but is most recognizable first in the smaller muscles, such as the facial muscles. Rigor mortis is another very important tool in determining the time of death.

After rigor mortis has fully set in, then another one or two days later, laxity, or the second flaccidity stage develops and the rigor disappears. Rigor occurs because there isn't any more oxygen in the body, but then the second flaccidity occurs as carbon dioxide builds in the muscle tissues, causing lactic acid to form and coagulate. As the lactic acid continues to form in the dead body, it becomes more acidic until this acid actually then dissolves the coagulant formed, and the muscles become flaccid again.

In hot climates, rigor begins one to two hours after death. However, in temperate climes, rigor does not begin for three to six hours: it takes an additional two to three hours to develop compared to hotter climates. Rigor comes on sooner in summer than in winter and lasts longer in winter. The reason is that the higher the ambient temperature, the sooner rigor begins, and the faster the stage progresses due to the heat, which provides a conducive environment for the metabolic processes in the body that cause decay. Lower ambient temperatures (i.e., temperatures of the surroundings) slow down the process of rigor. This is why if a person dies outside in a frozen environment, rigor may last several more days than if one dies in a warm environment. Interestingly, men tend to cool more quickly than women. Thin people cool more rapidly than heavy ones and children cool faster than adults. Also, obviously, a naked body will cool more quickly than a clothed body or a covered body, and a body in water cools much more rapidly than in air.

There are various techniques for estimating the time of death using rigor mortis. The atmospheric and environmental conditions are extremely important for an accurate determination of the time of death. In addition to ambient temperature, as already discussed, other parameters such as wind, rain, and snow are important clues. The windier it is, the more rapidly the body cools, and if the wind is cool, as compared to a warm wind (air), this also speeds up the cooling of the corpse. The body's temperature will cool until it reaches the ambient temperature, the temperature of the surroundings. If the ambient temperature is warmer than the body, the body will actually warm until it reaches the environmental temperature. This is because the body is trying to reach thermal equilibrium with the environment. In general, a corpse will lose 1.5 degrees F per hour until it reaches the ambient temperature. But this ambient temperature may change over time, especially if the body is outside. Also, a change in the environmental temperature of only 5 degrees F during the first twelve hours after death can create an error of plus or minus two hours in the timing of death. And if the body has been moved postmortem, potentially to a place with a different temperature, this adds another difficult wrinkle to the case.

One of the most common uses of forensic meteorology is to determine the length of time a person has been dead. This involves a meteorologist providing data for an entomologist (a "bug doctor") and/or a coroner. Why would an insect expert be involved? The reason is that as the body temperature cools and rigor sets in, the tissue (in both animals and humans) becomes attractive to a wide variety of insects and invertebrates, especially flies. The *Calliphoraidae*, the green and blue bottle flies, are usually the first suspects to arrive on scene. Then come the flesh flies (*Sarcophadigae*) as the body decays due to microbial fermentation. Then come the beetles and pyralid moths. Though many other species of insects and invertebrates may show up, this is the basic pattern followed around the globe.

Detailed entomological calculations take place in conjunction with forensic meteorology in order to determine the time of death of a body:

- **STEP 1.** Determine the temperature history at crime scene where the dead body is located.
- **STEP 2.** Rear maggots to adulthood to identify species: this involves collecting a range of maggots from the body, recording the time, and keeping the pupae until adult insects emerge.
- **STEP 3.** Estimate the time of egg-laying of the insects.
- **STEP 4.** Determine what other insect evidence is available.

All of the above depend on the weather conditions, including temperature and humidity.

But why would one require the services of a forensic meteorologist to aid in this determination when so much research has already been done on the specifics of how insects invade, lay eggs, and then grow in bodies? This is because it doesn't matter if you are the FBI, the NTSB, or the CIA—a forensic meteorologist is still the only one who can accurately determine what the atmospheric conditions were like during the entire time a particular body has been in its environment. Most dead bodies are not found directly next to a weather station. This means that a forensic weather detective must gather clues in order to determine the air temperature, relative humidity, sunlight (or lack thereof), and precipitation over the time the body has been exposed. Sometimes this can be days, weeks, or months. In some cases, the body has not been in the same place the entire time. In cases of foul play, where the murder takes place may not be where the body eventually winds up. This complicates things; the science becomes less certain and the possible time of death then spans a longer time frame due to uncertainties about whether was the body covered up at some point in time or whether it was outside in a different location prior to the location where it was found.

It was the afternoon of October 26, 1983, and the 21-year-old mother of an infant was reported missing by her parents. She was last seen alive that morning in the apartment of a male neighbor in Washington, D.C., where she resided. Witnesses said that they heard a woman screaming and saw a woman struggling with a man in the vicinity of the apartment building.

Eighteen days later, on November 13, a motorist traveling on a suburban Washington, D.C., highway pulled over on the side of the highway to pee in a wooded area about fifty feet from the roadside and came upon the body of a young, black female weighing about 116 pounds, which matched the description of the missing woman. Investigators arriving on the scene determined that the body lacked rigor mortis and was cold with minimal decomposition. The next day an autopsy was performed and the cause of death was revealed: multiple stab wounds to the chest and the neck. The stab wound in the chest penetrated the right ventricle of the heart and the stab wound in the neck perforated the esophagus and severed the right external carotid artery. During the autopsy a few large maggots were found. Some were found crawling away from the body. Apparently this is characteristic of fully developed blowfly larvae that are finished feeding. Other maggots were found in the wounds on the neck and in the corpse's clothing.

Investigators found hairs from the victim's head and pubic area on the bed sheets in the man's apartment where she was last seen. They also found one of her shoes in a wooded lot a short distance away from the apartment building where she was last seen. Bloodstains in the suspect's car matched the victim's blood and carpet fibers found on the victim were matched to the vehicle, so that they knew she had been transported in the suspect's car. Even with all the evidence, however, they needed an accurate time of death of the victim in order to establish a timeline of events to tie the suspect to the killing. Unfortunately, the autopsy did not reveal a time of death and none of the fly larvae remained alive to be reared to determine their exact stage in life. The crime scene vegetation and soil were examined, but they did not reveal any additional fly larvae or other evidence to pinpoint the time of death. The time of death as determined by the medical examiner and the investigators varied between two and eight days! There was too much margin of error, still. It was mandatory that they nail the timing down in order to link the suspect to the killing.

So it came down to the weather. The weather had be tied to the stage of the fly larvae found on the body, given that it appeared that the larvae were in their post-feeding stage, and one of the larvae was even the showing signs of pupation (beginning the pupal stage of metamorphosis). So information on the air temperatures, dew point temperatures (in order to determine relative humidity), cloud cover, precipitation, and winds (speeds, directions, and gusts) were gathered from the region. Fortuitously, there was an official National Weather Service observatory less than one-quarter of a mile from where the woman's body was found. Analyzing the weather data, especially the hourly air temperature data and the number of days to reach the "accumulated degree hours (ADH)" required for the *Calliphora vicina*, the blue bottle fly, to develop from egg to prepupa, it was determined that for the bottle fly larvae to reach the stage they were found on the body would have taken approximately *fifteen* days, as it had been relatively cool over that time period (average daily temperature of 50 degrees F). Taking this into account, it was eventually determined that the victim was murdered on the morning when the witnesses saw her struggling with the attacker.

Eighteen days later, her body was found. He killed her that morning and had hidden her body in a nearby urban wooded lot under tree branches, a mattress, and other debris. Three days later, he transported her body in his car to the location where it was eventually discovered by the motorist. While the body was covered when it was dumped, it was not yet infested with the fly larvae, so the combination of the weather data and entomology data from the flies was crucial in determining the exact time of death and thus confirming that that morning was indeed the last time she was seen alive.

The killer was convicted in a U.S. District Court of first degree murder, kidnapping, and felony rape, and sentenced to a lengthy prison term.

Death is never a pleasant topic, (unless your belief system supports some kind of paradise afterlife or a return ticket, coming back rich and beautiful). Plane crashes and death seem to go hand in hand. One of my plane cases was miraculous, however, given the condition of the aircraft after an accident on landing. Incredibly, there were very few fatalities. On August 22, 1999, Typhoon Sam was moving over the South China Sea toward Hong Kong. A typhoon, as you will recall, is the same thing as a hurricane or a tropical cyclone. Once the maximum sustained winds in a cyclone reach 74 miles per hour or higher, it is classified as either a hurricane, tropical cyclone, or a typhoon. These three terms represent the same phenomena; they are just classified differently depending on where in the world they originate.

As Typhoon Sam (called a typhoon because it was located in the northwest Pacific) made landfall near Hong Kong, it was downgraded to a severe tropical storm. It brought gale force winds, thunderstorms, and rain to the city. This storm brought more rainfall to Hong Kong than any other cyclone to this day since records began in 1884, with 616.5 millimeters or over 24 inches of rain, causing landslides and severe flooding in southeast China.

The typhoon made landfall around six P.M. in the evening in Hong Kong when it became a severe tropical storm. At this very time China Airlines flight 642 was headed from Bangkok, Thailand, to Taipei, Taiwan, via Hong Kong. At 6:06 P.M. the crew obtained weather information for the Hong Kong International Airport (HKIA) that confirmed that there was heavy rain, wind with gusty winds, and the visibility was less than a half a mile. Prior to the arrival of Flight 642, four flights carried out missed approaches (meaning they tried to get into the airport but couldn't because the visibility was too low) and five planes diverted to other locations where conditions were safer. All told, twelve flights landed successfully prior to the arrival of Flight 642.

At 6:41 P.M., Flight 642 was cleared to land. A video shot by passersby on the roadway along the airport shows the McDonnell Douglas MD-11 aircraft was encountering shifting winds as it attempted to land. The official findings were that two seconds prior to touchdown, the aircraft encountered a severe downdraft and the hard impact caused the right landing gear to collapse and the number three engine to hit the runway. The final report states that the right landing gear as well as the right wing were separated from the aircraft by the impact. The broken aircraft rolled and skidded down the runway in flames and ended up upside-down.

Of the 315 people on board, amazingly, only three people died. All fifteen crew members survived, but 219 people were taken to the hospital and fifty of these had severe injuries. The cause of the accident was determined to be pilot error, specifically the commander's inability to arrest the high rate of descent (sink rate) existing at fifty feet radar altimeter (i.e., fifty-feet above the ground). Just near touchdown, the aircraft was descending at eighteen to twenty feet per second (this is equivalent to 1,200 feet per minute, which is much too fast—500 feet per minute is a more normal descent rate for a plane approaching the runway). Contributing to the accident were the variations in the wind direction and speed just above the surface, thanks to the typhoon/tropical storm.

Working on this case for an attorney representing some of the plaintiffs, who were victims in the crash, I began gathering the weather data I needed to reconstruct the accident. Weather data availability varies greatly from country to country. In the United States we have our official weather data repository, the National Climatic Data Center. This government center is the world's largest repository of weather and climate data; its main data focus is, of course, on the United States, with the majority of the data available free of charge. Many other countries have extensive weather archives, but the majority of them charge fees for obtaining the data. Some of these fees can be astronomically high, so it is buyer beware: one must always ask for a quote before ordering anything. For this crash in Hong Kong I needed data from the Hong Kong Observatory, a department of the Hong Kong government that focuses on weather. I sent both a letter and an email requesting the data I required and waited. It was ten A.M. local time and the WeatherExtreme offices were buzzing with meteorologists and others working on various projects when the phone rang. "It's for you, Elizabeth," Steve Goates, our mapping expert, said through the open door of his office. "It is a lady calling about some data and that was all I could understand of what she said."

I picked up the phone. "Hello, may I help you?"

"Yes, doctor, you called for some data from Hong Kong," the lady replied with a thick Chinese accent.

"Oh, yes, I need some weather data for a specific time period," I responded.

Before I could say anything else, the Chinese lady asked what the data was for.

Knowing that she must have my information in front of her since she called me at the office, I knew she was fishing to find out if I wanted to reconstruct either the typhoon event or the plane crash. "I am reconstructing the weather events on a particular day when there was a typhoon that went through Hong Kong," I said, trying to avoid mentioning the accident. I wanted the data.

"One moment please," she said. While I was holding on, I could hear her rapidly speaking in Chinese and then I heard at least two other male voices responding to her in Chinese.

She was back on the line. "Why do you need data for August twenty-second? For what purpose?"

"I want to be able to accurately reconstruct the weather events on that day as the typhoon hit Hong Kong," I answered, hoping the language translation would be difficult enough that she would just give in. I heard her speaking in the background to the men again.

"Is this for the plane crash?" She got right to the point.

"Yes, it is to reconstruct the weather surrounding the accident," I said, knowing the cat was out of the bag now. She went back to conversing with the men.

"Are you an attorney?" she inquired.

"No, I am an atmospheric scientist," I said, hoping to put their fears to bed. But she continued her background conversation with the men.

"You work on a legal case for an attorney?" she asked, as they probed further into why I ordered the data.

"Yes, I have been retained by an attorney here in the United States to research the weather conditions," I replied, knowing it was up to them now. Another background powwow with the men and then she was back on the line.

"No data. We will give you no data," she retorted.

"But I am not an attorney and I only wish to analyze the weather conditions. I am not an accident reconstructionist, just a meteorologist," I pleaded, but to no avail. They had made up their minds and that was that. I thanked her and hung up the phone. She was obviously part of the same contingent that was doing the censoring in Harbin. Not much had changed in the past years since my visit there.

After the call, Steve Goates and I noticed that it must have been awfully late in Hong Kong for them to call. We converted the time and concluded that it was two A.M. local time in Hong Kong. How strange to think of these people all hovering around the phone at two A.M. just because a meteorologist in the United States ordered some weather data. But the stakes for the Chinese were high and this was to be a huge and important investigation. The Chinese government opted not to cooperate for fear of embarrassment from having made mistakes that resulted in disaster. In the end, I was able to obtain most of the data I required from other sources, including Australian, Japanese, and European weather centers, and successfully completed my work on the case. In the end an official report was issued and it has many recommendations that came out of this horrific accident, most of which had to do with either the China Airlines crew training and briefing, and recommendations that both the Boeing company and China Airlines should implement in terms of equipment and instruments on the airplane and on the airfield.

Hands-down my least favorite cases involve the death of children. The death of a child is horrible enough, but then to have to investigate the surroundings behind the fatality is sometimes more than I can bear. Especially when I find that the events that led up to the death were preventable!

The circumstances surrounding the deaths of children I have been involved with investigating range widely, from automobile accidents to the aforementioned parasailing accident. I do not look at places that are supposed to bring pleasure the same way I used to, especially the rides at county fairs and amusement parks.

One case still stays with me. The child was only eleven years old and hanging on for dear life hundreds of feet up in the air. It was in the evening and she had been alone on a ferris wheel. She had been seated by the ride attendant in a routine manner. It was windy that day, as had been forecast. This meant that when the winds became too gusty, the person in charge of the wheel had to stop it, wait for the gusts to cease, and then restart it. But when the wheel stops, it causes the chairs to swing along with the wind still pushing the chairs. The ferris wheel in question was 160 feet high at the top. The winds were gusting and shifting direction. At the time there was a direct crosswind with respect to the wheel. In general, wind speeds increase with altitude and on this day the data from surrounding stations and weather balloons show that was indeed the case. How easy it would have been to have at least one anemometer (device for measuring wind speed), even a hand-held one, for the employees working at the base of the ferris wheel to know when to stop the ride. Even better, mount one on top of the wheel to get an idea of how much gustier it is up there. It could be checked with every rotation of the wheel.

The little girl who was riding alone somehow slipped out of her chair, perhaps in a panic at being stopped so high up with the winds gusting about her. She held on for a while, but eventually could not hold on any longer. She was near the very top of the 160-foot-tall wheel. All that lay below was concrete, metal, and wood. She landed in the passenger loading area and did not survive.

These are the cases that never leave me—the ones that could have been preventable if only proper heed had been paid to the weather.

NINETEEN

"Saint Francine" of Assisi

Eist le mo chroí
As é leanúint ailsing'
As é ag gabháil go háit
Nach eol ach dom féin
Ní heol do mo chroí
Ach an aisling a leanúint
Nach Féidir ach liomsa a haithint

Listen to my heart
As it follows a dream
And it leads to a place
That only I know
To know my heart
Is to follow a dream
A dream that only I can know

—Patrick Moody Williams, from the song
"A Dream That Only I Can Know" (Gaelic and English)

These lyrics are from a beautiful song that my dad wrote; they still resonate with me. Sometime back, one of my dad's friends was telling him about his son who was in college and interested in music. He said to my father, "He is trying to decide if he wants to be a business major or a music major." My dad told me he responded, "If he has to make a decision, he should choose business." I often wonder if the man understood what my dad was really saying. If you are not drawn with all of your heart and soul to something, don't do it.

My husband, Alan, lovingly dubbed me "Saint Francine of Assisi" years ago. I have always loved, admired, and respected animals and the environment. Animals have always been and will always be a very important part of my life. When I was a child, we always had animals. Lots of animals. We had snakes, birds, dogs, cats, horses, turtles, hamsters, bunnies, rats, lizards . . . you name it, we had it. If anyone in the neighborhood found an injured or lost animal, wild or domesticated, they called the Williamses. One day, a man named Clark Terry, who was a world-renowned jazz trumpeter, came to our house to visit with my dad. He and my dad had been friends since before I was born. As he walked through the living room to get to the back of the house where my dad's office was, he saw lying on the living room sofa our dog Noodle and our cat Pebbles. Noodle was by then being called "Mother Noodle," as she had had a few litters of puppies over the years. When Clark saw her, she was about six pounds and was *nursing* our orange tabby cat, Pebbles, who weighed about twenty pounds. Clark Terry took one look at them and blurted out, "THAT belongs in a ZOO!" Our family still laughs about it to this day.

"I always know where you are," Alan said to me, "as all I have to do is to look for a trail of pets. They follow you everywhere and never leave your side."

"I know, and I love it. They must feel my love for them," I responded, knowing full well I am no different than any other pet owner who feels so special due to their pets' love and affection. Birds, plants, mammals, fish, you name it: they are all amazing creatures on this planet. I am a firm believer that we humans have a duty to ensure that we have and maintain a healthy planet for ourselves and for these wonderful creatures to live on and to continue living on.

Alan and I have worked together for years as *yin* and *yang*, in the best sense, by always complementing and supporting each other. Together, as husband and wife, as friends, as parents, and as partners in WeatherExtreme Ltd., we have always held the belief that a successful company is only as good as its employees. By having employees that we also like as people, we have achieved this in our company. That is also why we have always included family as part of our business. My brother, Patrick, has been an integral part of the company for years. He has his master's in fine arts from Carnegie Mellon University and is in charge of all of our graphics.

I was sitting on the back porch of our home reminiscing about friends, family, work, and just life itself. I was thinking back, realizing that I founded the company more than twenty years ago. At the time I started the company, little did I realize that I would get to work on the NASA X-43A Hypersonic Research Vehicle project; participate in the Perlan Project's attempt to soar a glider to a record 100,000 feet; work on more than 1,500 forensic meteorology cases; and travel to six of the seven continents and visit over 15 countries just for my work alone! Some of the plane crash cases I have worked on include the Air France Flight 447 crash; Governor Mel Carnahan's plane crash in Missouri; Senator Ted Stevens' plane crash in Alaska; the crash that killed U.S. Secretary of Commerce Ron Brown; New York Yankees pitcher Cory Lidle's crash; the Oklahoma State University basketball team crash; and the crash of singer Aaliyah's plane in the Bahamas. Many of the accidents I have worked on have been due to weather conditions; some were directly attributable to pilot error; and some were a combination thereof. Along the way, I discovered new facts about clouds, precipitation, winter mountain storms, stratospheric mountain waves, and climate. Also along the way I met and came to know so many wonderful people. One cannot put a price tag on these experiences.

As I reflected back over the past twenty years, the famous theoretical physicist, Dr. Wolfgang Pauli, who was awarded the Nobel Prize in Physics in 1945 for his discoveries in the field of quantum physics, came to my mind. Pauli wrote an essay about the scientific theories of Johannes Kepler and his discovery of three major laws of planetary motion. Pauli pontificated about Kepler's process of scientific discovery and aptly says that, though scientific theories themselves are described by mathematical functions and formulas, the actual process of discovering a new scientific theory is quite irrational. One must come up with these ideas through creativity and risk. Pauli states, "The only acceptable point of view appears to be the one that recognizes *both* sides of reality—the quantitative, the physical and the psychical—as compatible with each other, and can embrace them simultaneously." Pauli often wrote about the psychology and philosophy of physics and of trying to understand the universe. "The best that most of us can hope to achieve in physics is simply to misunderstand at a deeper level," he wrote to physicist Dr. Jagdish Mehra. Pauli felt that his philosophy of things was similar to Einstein's in that there is a central order of things and their shared belief in the simplicity of natural laws.

Climate change, pollution, extreme weather events, and environmental problems can be addressed—and solutions achieved—by the collaboration of public and private entities, from small to large institutions, to governments and the general public. It will take young and old, scientists and non-scientists, to achieve our goals. I fondly remember my discussions with other scientists over the years about the "Gaia Hypothesis" and how we each felt about it. Gaia is an ancient Greek goddess of Earth, complete in herself. The Gaia goddess was "the first to arise from chaos," as described in Hesiod's poem "Theogony" from 700 BC. This ancient Greek's poem describes the origin of the cosmos and the relationships of the gods of the ancient Greeks. But long before that, 1,300 years before, Gaia (or "Gaya") was the first hymn to arise from the original seed sound of Om, and in Sanskrit, the classical language of India, it meant "Moving Song." The Gaia Hypothesis was proposed by Dr. James Lovelock, a British chemist, back in 1965; this hypothesis proposes that the climate system (atmosphere, cryosphere, hydrosphere, lithosphere, and the biosphere) is one complex interacting system that maintains conditions on Earth, and this system tries to remain in a state of homeostasis. What exactly does that mean? It means that the Earth should be considered as a single living system or organism that is alive and always striving to maintain relatively constant or stable conditions. That is, the system—our Earth—self-regulates. The longer I study the atmosphere, the more I adopt this philosophy. The atmosphere and the ocean are interconnected. Remember, "energy cannot be created or destroyed, it can only be changed from one form to another," (as Albert Einstein said) and the universe is infinite. We as humans are connected to the atmosphere and the ocean. We are also connected to the rest of the biosphere, and so on. So we are one complex, yet connected, single system. If we damage one part of the system, we are only hurting ourselves.

To take a simplistic, but, I think, effective, look into the Gaia Hypothesis, there are many examples over the years how Mother Nature fools us and winds up reacting in ways we didn't see coming. She self-regulates. Remember another thing Einstein said: "Two things are infinite: the universe and human stupidity." The kudzu plant was brought to the United States from Japan and China. It is a perennial vine native to eastern Asia and some Pacific islands. Kudzu was introduced to the U.S. in 1876 as an ornamental plant, and in the 1930s it was planted all around the southern United States by agricultural officials to control erosion on hillsides. Kudzu grows at an amazing rate and, more important, is a parasite: it grows at the expense of other plants, including native species, by taking their space and sunlight. Now it is a massive pest that has destroyed local plant life, and people continually struggle to control it.

Introducing new species to a foreign environment usually has unexpected outcomes. The purple loosestrife was planted in gardens in the United States because it produces beautiful flowers. This wetland plant was first brought to the East Coast of the U.S. in the 1800s. It is native to Europe and Asia. These plants now grow in forty of the U.S. states and in all Canadian provinces that border the U.S. Being a wetland plant, this invasive species now grows along lakeshores, ponds, rivers, and wetlands. This plant grows so that it forms dense cover that is then impenetrable, so that it is not suitable for cover for such creatures as ducks, geese, frogs, even muskrats. Muskrats are medium- to large-sized rodents that have a furless tail and webbed back feet. They live in marshes or along waterways. They eat a variety of food but their favorite food is the cattail plant. The cattail is a tall, reed-like plant found in wetlands. Muskrats also use these plants to build their nests. In places where the purple loosestrife have taken over and pushed out the cattails, the muskrats disappear. The muskrats hasten the invasion of the purple loosestrife by eating their competitor, the cattail. This reduction in cattails has also reduced the number of water birds, which then reduced the numbers of the predator that eats these birds, the mink. Soon, an entire ecosystem has vanished.

These situations show how nature winds up responding to outside inputs in ways we didn't consider. Though we better understand the impact of invasive flora and fauna now, we are still seeing consequences to our actions that we did not expect—or chose to ignore. That is what is happening to our atmosphere and our oceans. We are overpopulating our planet and are deluging it with particulates, chemicals, and toxins. We are also changing the atmosphere through deforestation, the construction of new cities and towns, and the paving over of raw land. These changes in the land have a dramatic effect on the weather and climate. Obviously not everything we are doing in the name of progress is bad for the environment, but it *is* an unknown, as is how we are changing our environment. All we know is that we *are* changing our environment. We just don't yet know the full spectrum of consequences. But what we *do* know is that the global surface temperatures are rising, the temperatures in the troposphere are rising, the stratosphere is cooling, and the oceans are heating up. How the environment and ecosystems will handle these changes in the long term is not known. Yes, we know what is occurring now and can scientifically postulate what may occur in the future, but we do not know for sure. This is where the Gaia Hypothesis comes into play for me. The climate system works as a whole and thus responds as a whole. When we impact one portion of it, we are impacting the whole of it. This can be exciting for scientists and for all human beings as we discover new things, but it can also be terrifying, as there is so much unknown. But we can all do something about our changing climate. It is not too late. Not everyone can do everything, but if we work together as a whole, we can reduce pollutants and the amount of greenhouse gases we put in our atmosphere, we can reduce our waste, we can recycle some of our waste, we can look for alternatives to deforestation, and we can educate everyone, especially the younger generations, on how to care for our environment and our atmosphere. As has been stated by many pundits and regularly discussed in the media, small changes even in your own home, small business, or school combine to make a large effect.

If we humans each lived for four hundred years, would we live our lives differently? We would need to live with the consequences of what we are doing now and thus might behave differently. These are the types of questions we need to ask ourselves. We as humans are not the deciders on the planet, we are only caretakers, as Dr. Lemuel Gulliver, the protagonist in Jonathan Swift's novel *Gulliver's Travels*, found out. Gulliver, when in the land of the Brobdingnagians, the sixty-foot-tall people, was quickly captured and presented to the king. The king of Brobdingnag uttered words that still make me shiver, as written by Larry Gelbart in *Gulliver*. These are the thought processes that need to be eradicated for the environmental problems we face to be tackled.

> *Word of the magical creature spread through the Kingdom,*
> *and soon there appeared a royal emissary,*
> *sent by her Mammoth Majesty, The Queen of Brobdingnag.*
> *Pieces of gold changed hands; the farmer becoming rich*
> *enough to buy his pigs the finest of slop,*
> *while the gift-wrapped Gulliver was presented*
> *to the highest of all highnesses, the King of Brobdingnag,*
> *who opined at once that human nature seemed a*
> *contemptible thing when it could be mimicked by a*
> *bite-sized button such as Gulliver.*

Although it is perhaps not their nature, I believe that scientists must step up to the plate and play a larger role with respect to educating the media, politicians, and the so-called giants of industry. Remember, these people are *not* scientists, and it is the responsibility of scientists to ensure that the information being spread about our climate and environment is accurate. We cannot pay attention to phony findings that are used to put a spin on a story to get ratings, votes, or some additional cash for some greedy conglomerate's bank account. And it is our duty as voters and consumers to make sure that neither the media nor the politicians nor industry prevent true scientists from being able to get the message out.

I look at scientists as life's expert witnesses. An expert witness is different than a lay witness in court. Lay witnesses in court can testify to what they saw or what they know and their perceptions, but these are not based on "scientific, technical, or other specialized knowledge." An expert witness, on the other hand, is held to much higher standards. In fact, here is the main portion of the U.S. Federal Rules of Evidence, namely Rule 702, for Testimony by Expert Witnesses:

> *A witness who is qualified as an expert by knowledge, skill, experience, training, or education may testify in the form of an opinion or otherwise if:*
>
> *(a) the expert's scientific, technical, or other specialized knowledge will help the trier of fact to understand the evidence or to determine a fact in issue;*
> *(b) the testimony is based on sufficient facts or data;*
> *(c) the testimony is the product of reliable principles and methods; and*
> *(d) the expert has reliably applied the principles and methods to the facts of the case.*

Just as important is the fact that experts must only testify to facts of their field; they cannot stray out of their area of expertise. I would not want to hear factual details about how a plane is operated and flown from an ophthalmologist, just as I wouldn't want a pilot performing eye surgery on me. In addition, the Federal Advisory Committee made notes on these rules. These notes state the following:

> *Whether the situation is a proper one for the use of expert testimony is to be determined on the basis of assisting the trier. "There is no more certain test for determining when experts may be used than the common sense inquiry whether the untrained layman would be qualified to determine intelligently and to the best possible degree the particular issue without enlightenment from those having a specialized understanding of the subject involved in the dispute." When opinions are excluded, it is because they are unhelpful and therefore superfluous and a waste of time.*

I find this "test" for determining if one is a qualified expert useful in sorting through the reams of articles and essays written about the atmosphere, the weather, and climate change. I look at from whence it came, the source of the information. If it does not have a credible source, I can better see it for what it is—a spin story or biased finding meant to further profits or achieve some other goal. Anything but real science.

I have such fond memories of watching Jacques Cousteau and his underwater adventures on television and reading his books as a child and as an adult. He opened my eyes to the amazing underwater world on this planet. Now Jean-Michel Cousteau, one of Jacques's four children, is continuing on in his father's footsteps. He is a tremendous advocate for the oceans and the environment. In an interview he was asked, "Do you think that these economic goals for improved science education are in conflict with the environmental goals that you cite?" He responded, "No, they're not. I really believe that science is critical, and science and the environment are really one and the same thing. We need to take care of Planet Earth, and our beautiful planet is in distress due to the terrible lack of awareness and mismanagement of our resources on so many levels. Hopefully anyone who has a real understanding of the situation will consider adapting and changing. But how can you protect what you don't understand? Here is where education is critical because that's when the future decision makers are acquiring the information which will allow them to make better decisions when they become adults."

The future for our planet is bright but only if we act *now*. Alan, Evan, and I recently watched a documentary on garbage in our oceans. After watching that film, our son began watching videos on his own on the Internet about our planet and pollution. He grabbed me one day and took me into the bathroom and began pointing out all of the bottles of body spray, hair spray, toothpaste tubes, and toothbrushes. As he pointed out each one, he said to me, "That one is made out of plastic, and that one and that one . . ."—on and on. Then he took me into his room and pointed out all of the items and what they were made of, some of wood and metal, like his bed, but many more plastic items. It opened up his eyes to how much plastic we have on the planet. Now, when we go to the grocery store together and they ask, "paper or plastic?" I turn to Evan and let him answer. "Paper, please," he announces. But paper takes trees, water, and energy to create, so he now wants to make sure we use our own reuseable bags. Now this may seem sophomoric, but these little steps lead to big steps. This is how changes are made. But we cannot dictate how others live. We all have our own personal beliefs and opinions on how we live our lives. Some people don't eat meat, some people live in tents, others in mansions, while others don't eat much at all and their only concern is food and water. Others are living with diseases or chronic illness. We cannot expect everyone to follow the same path. But we can expect that we all follow a path that leads to betterment of the planet and our environment. We can expect this *and* demand it of our politicians, the media, and ourselves.

A few years ago, Alan and I were meeting Mrs. Ware for the first time. She was to be Evan's second grade teacher. My husband and I listened as Evan announced to her, with great pride, "My mommy is a scientist!"

I had definitely *listened to my heart.*

Quod Erat Demonstrandum (Q.E.D.)

Acknowledgments

I am indebted to my loving family, not all whom are still here on earth with us. I am indebted to Mr. Dan Strone, CEO of Trident Media, my literary agent and confidant. Many thanks to Pegasus Books and their belief in me and my book . . . especially Jessica Case, for her patience as a publisher and editor.

I also wish to acknowledge a few of many people for their contributions to this book. These are: Dr. Ashley Smart, the American Institute of Physics, the American Meteorological Society, and their Certified Consulting Meteorologists (CCM) program.

I owe a debt of gratitude to many people along the way. My dear friends Einar Enevoldson and his wonderful wife, Susana Conde. Steve Fossett, who is dearly missed, and the Perlan Project lives on in his memory. Gene Schwam for his guidance and wisdom. My lifelong friends and cohorts, Dr. Belay Demoz and Ed Teets, Jr.

Selected References/Suggested Reading/Suggested Listening

CHAPTER 1

Austin, E. J. "The Perlan Project Phase II: Soaring Stratospheric Mountain Waves in Argentina to 90,000 feet." 28th Conference on Interactive Information Processing Systems: International Applications, 92nd AMS Annual Meeting January 21–26, 2012, New Orleans, Louisiana.

Carter (Austin), E. J. and E. H. Teets, Jr. "Stratospheric Mountain Waves: Observations and Modeling for a Proposed Sailplane That Will Use These Waves to Reach 100,000 Feet." Preprint in 18th Conference on Interactive Information Processing Systems (IIPS) for Meteorology, Oceanography, and Hydrology, 82nd Annual American Meteorological Society Meeting, January 13–17, 2002, Orlando, Florida, 279–281.

Dubois, T. "Airbus-sponsored Perlan 2 to soar to 100,000 feet in bid to understand climate." *Aviation International News*, November 5, 2015.

Durran, Dale R. "Lee Waves and Mountain Waves." in the *Encyclopedia of Atmospheric Sciences*, 2003, Holton, J. R., J. Pyle, and J. A. Curry, eds., Elsevier Science Ltd., 1161–1169. With permission from Elsevier.

Eckerman, S. D. et al. "Mountain Waves in the Stratosphere." *Naval Research Laboratory Review*, NRUPun641-00-411, Washington, D.C., June 2000.

Krishnamurti, T. N. "A vertical cross section through the 'polar-night' jet stream." *Journal of Geophysical Research Atmospheres*, 1959, 64(11): 1835–1844.
Labitzke, Karin G. and Harry van Loon. *The Stratosphere: Phenomena, History and Relevance*. Berlin, Heidelberg: Springer-Verlag, 1999.
Lester, Peter F. *Turbulence: A New Perspective for Pilots*. Inverness, Colo.: Jeppesen Sanderson, Inc., 1994.
NASA Goddard Space Flight Center. *What is the Polar Vortex?* http://ozonewatch.gsfc.nasa.gov/facts/vortex_NH.html
The Perlan Project. http://www.perlanproject.org
Reynolds, James. *The Hidden Dangers of Mountain Wave Turbulence*. NOAA's National Weather Service, *The Front*, Nov. 2011.
Strauss, L., S. Serafin, S. Haimov, and V. Grubišić. "Turbulence in breaking mountain waves and atmospheric rotors estimated from airborne *in situ* and Doppler radar measurements." *Quarterly Journal of the Royal Meteorological Society* Vol. 141, 693 (2015):3207–3225.
Whelan, R. F. *Exploring the Monster. Mountain Lee Waves: The Aerial Elevator.* Stockton, Calif.: Wind Canyon Books, 2000.

CHAPTER 2

American Heart Association, "Cold Weather and Cardiovascular Disease." http://www.heart.org/HEARTORG/General/Cold-Weather-and-Cardiovascular-Disease_UCM_315615_Article.jsp#.Vpa88MArI18
British Heart Foundation, "Wise Up To Winter." https://www.bhf.org.uk/heart-health/living-with-a-heart-condition/cold-weather
Flood Insurance. Insurance Information Institute (www.iii.org), November 2015.
Geiger, Peter, Philom and Sondra Duncan. *Farmer's Almanac 2016*, Lewiston, Me.: Almanac Publishing Company, 2015.
Harkavy, J. "Farmer's Almanac Predicts Very Cold Winter." Associated Press, August 24, 2008.
Hauser, Rachel. "Location, Location, Location." NASA Earth Observatory Feature Article, August 15, 2001. http://earthobservatory.nasa.gov/Features/Location/
Hsiang, S. M., M. Burke, and E. Miguel. "Quantifying the Influence of Climate on Human Conflict." *Science*, Vol. 321, No. 6151, September 13, 2013, DOI: 10.1126/1235367.
Oxley, Lawrence J. *Extreme Weather and Financial Markets: Opportunities in Commodities and Futures*. Hoboken, N.J.: John Wiley & Sons, Inc., 2012.
Sachs, Jeffrey D., Andrew D. Mellinger, and John L. Gallup. 2001. "The Geography of Poverty and Wealth." *Scientific American*, March 2001.
Starr-McCluer, M. *The Effects of Weather on Retail Sales*. Washington, D.C.: Federal Reserve Board of Governors, 2000.
Stillman, Janice, ed. *The Old Farmer's Almanac 2016*. Dublin, N.H.: Old Farmer's Almanac, 2015.
Walsh, J. E. and D. Allen. "Testing the Farmer's Almanac." *Weatherwise*, 34 (1981): 212–215.
Weather Ready Nation. National Oceanic & Atmospheric Administration. http://www.nws.noaa.gov/com/weatherreadynation/
WeatherExtreme Ltd., http://www.weatherextreme.com

CHAPTER 3

Austin, Elizabeth and Peter Hildebrand. "The Art and Science of Forensic Meteorology." *Physics Today*, 67(6), 32, June 2014.

Baum, Marsha L. *When Nature Strikes: Weather Disasters and the Law.* Santa Barbara, Calif.: Praeger Publishers, 2007.

Cohn, Ellen G. "Weather and Crime." *British Journal of Criminology*, Vol. 30, No. 1, Winter 1990.

McLaren, C., J. Null, J. Quinn. "Heat Stress from Enclosed Vehicles: Moderate Ambient Temperatures Cause Significant Temperature Rise in Enclosed Vehicles." *Pediatrics*, 116 (2005): 109–112.

Null, Jan. "A Stationary Danger." *Weatherwise*, 58(4), July/August 2005.

Ransom, Matthew. "Crime, Weather and Climate Change." *Journal of Environmental Economics and Management*, Vol. 67, No. 3, May 2014, pp. 274–302.

Santa Ana. California Nevada Applications Program/California Climate Change Center. http://meteora.ucsd.edu/cap/santa_ana.html

CHAPTER 4

All About History of Weather and The National Weather Service. National Weather Service/NOAA. http://www.nwas.org/links/history.php

Aristotle. *Meteorologica.* Translated by H.D.P. Lee. Cambridge, Mass.: Harvard University Press, 1952.

Hall, R. Cargill. *A History of the Military Polar Orbiting Meteorological Satellite Program.* Office of the Historian, National Reconnaissance Office, Center for the Study of National Reconnaissance. September 2011.

Heise, René, Manfred Reinhardt, and Peter F. Selinger. *Dr. Joachim P. Küttner: Aeronautical Pioneer, Record Pilot, Meteorologist.* Translation: Elke Fuglsang-Petersen. *SegelFliegen International* 1 (2012): 56–59.

Hicks, J. W. *Flight Testing of Air Breathing Hypersonic Vehicles.* NASA Technical Memorandum 4524, October 1993.

NASA Armstrong Fact Sheet: Hyper-X Program. Feb. 28, 2014. http://www.nasa.gov/centers/armstrong/news/FactSheets/FS-040-DFRC.html

Van Wie, D. M., S. M. D'Alessio and M. E. White. "Hypersonic Airbreathing Propulsion." *Johns Hopkins APL Technical Digest*, Vol. 26 (2005), 4: 430–437.

CHAPTER 5

Cloud Appreciation Society. https://cloudappreciationsociety.org

Day, John A. and Vincent J. Schaefer. *Peterson First Guide to Clouds and Weather.* Boston: Houghton Mifflin Company, 1991.

Hamblyn, Richard, *Extraordinary Clouds: Skies of the Unexpected from the Beautiful to the Bizarre.* Devon, England: David & Charles, in association with the Met Office, 2009.

Hill, Jerry D. and Gerald J. Mulvey. "The Ethics of Defining a Professional: Who is a Meteorologist?" *Bulletin of the American Meteorological Society*, American Meteorological Society, July 2012.

Klosterman, Chuck. "Forecasting Fraud." *New York Times Magazine*, April 21, 2013.

Mandelbrot, Benoît B. *The Fractal Geometry of Nature.* New York: W.H. Freeman and Company, 1977.

Pretor-Pinney, Gavin. *The Cloud Collector's Handbook*. San Francisco: Chronicle Books, 2011.
Schaefer, Vincent and John A. Day. *A Field Guide to the Atmosphere*. Boston: Houghton Mifflin Company, 1981.
Shawe-Taylor, Desmond and Jennifer Scott. *Dutch Landscapes*. Berkshire, England: Royal Collection Trust, 2010.
Williams, Jack. *The AMS Weather Book: The Ultimate Guide to America's Weather*. Chicago: University of Chicago Press and the American Meteorological Society, 2009.

CHAPTER 6

Carter (Austin), E. J. and R. D. Borys. "Aerosol-Cloud Chemical Fractionation: Analysis of Cloud Water." *Journal of Atmospheric Chemistry*, 17 (1993): 277–292.
Hindman, E. E., E. J. Carter (Austin), R. D. Borys, and D. L. Mitchell. "Collecting Supercooled Droplets as a Function of Droplet Size." *Journal of Atmospheric and Oceanic Technology*, 9 (1992), 4: 337–353.
Hindman, Edward E. "An Undergraduate Field Course in Meteorology and Atmospheric Chemistry." Snow and Glacier Hydrology, Proceedings of the Kathmandu Symposium, November 1992, IAHS Publ. No., 1993, pp. 59–65.
Libbrecht, Kenneth and photography by Patricia Rasmussen. *The Snowflake: Winter's Secret Beauty*. London: Voyageur Press, 2003.
Mitchell, D. L. and R. D. Borys. "A field instrument for examining in-cloud scavenging mechanisms by snow." In *Precipitation Scavenging and Atmosphere-Surface Exchange*, Vol. 1: Precipitation Scavenging Processes, ed. S. E. Schwartz and W. G. N. Slinn, Boca Raton, Fla.: CRC Press, 239–253. 1992.
Rogers, R. R. *A Short Course in Cloud Physics*, second edition. Oxford: Pergamon Press Ltd., 1979.
Stohl, Andreas, editor. "Inter-continental Transport of Air Pollution." *The Handbook of Environmental Chemistry 4G*, London: Springer, 2004.
Storm Peak Laboratory, Desert Research Institute, University of Nevada, Reno. http://stormpeak.dri.edu

CHAPTER 7

"SAGE: The New Aerial Defense System of the United States." *The Military Engineer*. March–April 1956.
Leading the Way: The History of Air Force Civil Engineers 1907–2012. Ronald B. Hartzer, et al. of R. Christopher Goodwin & Associates, Inc.
Ahrens, Donald. *Essentials of Meteorology: An Invitation to the Atmosphere*, 7th Edition. Boston: Cengage Learning, 2014.
Beneath Northern Skies: An Account of Research Carried Out at High Latitudes, 1950 to 1959. Kinsey Anderson, 2001.
Burt, Stephen. *The Weather Observer's Handbook*. Cambridge: Cambridge University Press, 2012.
Desert Research Institute, Environmental Research Arm of the University of Nevada Higher Education System, http://www.dri.edu/
Fauve, M. and H. U. Rhyner. "Physical Description of the Snowmaking Process Using the Jet Technique and Properties of the Produced Snow." in *Snow Engineering V*,

edited by P. Bartelt, et al., Proceedings of the Fifth International Conference on Snow Engineering, July 5–8, 2004, Davos, Switzerland.

Grubišić, Vanda and John M. Lewis. "Sierra Wave Project Revisited: 50 Years Later." *Bulletin of the American Meteorological Society, AMS*, August 2004, 1127–1142.

Lawrence, Mark G. "The Relationship Between Relative Humidity and the Dewpoint Temperature in Moist Air: A Simple Conversion and Applications." *Bulletin of the American Meteorological Society, AMS*, February 2005, 225–233.

Male, D. H. and D. M. Gray. *Handbook of Snow: Principles, Processes, Management of Use*. Caldwell, N.J.: Blackburn Press, 1981.

Mohanakumar, K. *Stratosphere Troposphere Interactions: An Introduction*. Springer Science + Business Media B.V., 2008.

Mozer, Forrest. *Kinsey A. Anderson 1926–2012: A Biographical Memoir.* Washington, D.C.: National Academy of Sciences, 2014.

CHAPTER 8

Ackerman, Steve. *Contrails*. Space Science & Engineering Center, University of Wisconsin. http://cimss.ssec.wisc.edu/wxwise/class/contrail.html

Aircraft Contrails Factsheet, U.S. Environmental Protection Agency, Air & Radiation (6205J), EPA430-F-00-005, September 2000.

Bomar, George W. "Weather Modification and the Law." *Southwest Hydrology*, March/ April 2007, 22–23.

Carter (Austin), E. J. and A. B. Long. "The Water Budget and Precipitation Efficiency of Australian Winter Mountain Storm Clouds." Seventh Conference on Weather Modification, Chiang Mai, Thailand, 1999: Vol. I, 105–108.

Dennis, Arnett S. *Weather Modification by Cloud Seeding*. Cambridge: Academic Press, Inc., 1980.

Dennis, Arnett. "Cloud Seeding and the Rapid City Flood of 1972." *Journal of Weather Modification*, 42, April 2010: 124–126.

Great Lakes Storm November 9–11, 1998: Edmund Fitzgerald Remembered. Kirk Lombardy, Marine Forecaster, NWS, Cleveland, Ohio.

Hultquist, T. R., M. R. Dutter, and D. J. Schwab. "Reexamination of the 9–10 November 1975 'Edmund Fitzgerald' Storm Using Today's Technology." *Bulletin of the American Meteorological Society*, May 2006, 607–622.

Langmuir, Irving. *The Collected Works of Irving Langmuir*. Volume 11: Cloud Nucleation. C. Guy Suits, editor. Oxford: Elsevier Science, Pergamon Press, 2013.

Lightfoot, Gordon, "The Wreck of the Edmund Fitzgerald," *Gordon Lightfoot Complete Greatest Hits*, Warner Bros. Records, Inc. & Rhino Entertainment Company, 2002.

Lightfoot, Gordon, "The Wreck of the Edmund Fitzgerald," lyrics published by WB Music Corp.

Long, A. B. and E. J. Carter (Austin). "Australian Winter Mountain Storm Clouds: Precipitation Augmentation Potential." *Journal of Applied Meteorology*, 35 (1996): 1457-1464.

Long, Alexis B. "Review of Persistence Effects of Silver Iodide Cloud Seeding." *Journal of Weather Modification*, 33, 1, 2001.

"Nkosi Sikilele I'Africa" (South African National Anthem) Zulu Version. Nkosi Sikilele I'Afrika/Die Stem, Imilonji Kanty Choral Society, Gallo Music Productions, 1994.

Oroville, Harold D. et al. "A Response by the Weather Modification Association to the National Research Council's Report Titled *Critical Issues in Weather Modification Issues.*" The Report of a Review Panel, Weather Modification Association, January 2004.

Paton, Alan. *Cry, the Beloved Country*. New York: Scribner, 1948.

Schwarz, F. K., L. A. Hughes, and E. M. Hansen. *The Black Hills–Rapid City Flood of June 9–10, 1972: A Description of the Storm and Flood.* A report prepared by the United States Geological Survey and the National Oceanic and Atmospheric Administration, Geological Survey Professional Paper 877, U.S. Government Printing Office, 1975.

U.S. Geological Survey, *Desalination: Drink a Cup of Seawater?* http://water.usgs.gov/edu/drinkseawater.html

Van Sebille, Erik. "The Oceans' Accumulating Plastic Garbage." *Physics Today*, February 2015, 60–61.

CHAPTER 9

Baray, Jean-Luc, D. Ancellet, and B. Legras. "Planetary-scale Tropopause Folds in the Southern Subtropics." *Geophysical Research Letters*, 27 (2000), No. 3, 353–356.

Carillet, Jean-Bernard and Brandon Presser. *Lonely Planet Mauritius, Reunion & Seychelles (Multi Country Travel Guide)*. 7th Edition, Lonely Planet, 2010.

King, David A. *World Maps for Finding the Direction and Distance to Mecca: Innovation and Tradition in Islamic Science.* London: Brill, 1999.

NASA Goddard Space Flight Center. *Ozone Hole Watch*, http://ozonewatch.gsfc.nasa.gov

U.S. Environmental Protection Agency. *Ozone: Good Up High, Bad Nearby.* http://www3.epa.gov/airnow/gooduphigh/ozone.pdf

CHAPTER 10

Blum, Arlene. *Annapurna: A Woman's Place.* San Francisco: Sierra Club Books, 1980.

Guyer, J. Paul. *Ethical Issues from the Tacoma Narrows Bridge Collapse.* CreateSpace Independent Publishing Platform, 2013.

Hazari, Zahara, Philip M. Sadler, and Robert H. Tai. "Gender Differences in the High School and Affective Experiences of Introductory College Physics Students." In *Women in Physics: A Collection of Reprints in Honor of Melba Newell Phillips*, edited by Jill Marshall. College Park, Md.: American Association of Physics Teachers, 2014.

Hobbs, Richard S. *Catastrophe to Triumph: Bridges of the Tacoma Narrows.* Pullman, Wash.: Washington State University Press, 2006.

Intriligator, Devrie S. "Seventh Annual NOAA-Industry Space Weather Summit Meets." *Space Weather*, 11 (2013): 545–546.

Ivie, Rachel, et al. *Women in Physics & Astronomy Faculty Positions: Results from the 2010 Survey of Physics Degree-Granting Departments.* American Institute of Physics Statistical Research Center, American Institute of Physics, August 2013.

Moskowitz, Clara. "Why Women Don't Pursue Physics Careers." *Live Science*, April 3, 2012.

Petroski, Henry. *To Engineer is Human: The Role of Failure in Successful Design*. New York: Vintage Books, 1992.
Potvin, Geoff and Robert H. Tai. "The Relationship Between Doctoral Completion Time, Gender, and Future Salary Prospects for Physical Scientists." *Journal of Chemical Education*, 03(2012); 89(1).
Sierra Nevada College, Incline Village, Nevada. http://www.sierranevada.edu/
Space Weather, Carmel Research Center, Dr. Devrie S. Intriligator, Director. http://www.carmelresearchcenter.com/spaceweather/
Tacoma Narrows Bridge Collapse, video from the Prelinger Archives, San Francisco, Calif. https://archive.org/details/Tacoma-Narrows_Bridge_Collapse

CHAPTER 11

Barber, Samuel. "Adagio for Strings," conducted by Leonard Bernstein and performed by the New York Philharmonic Orchestra, from the album *Barber's Adagio and Other Romantic Favorites for Strings*, Sony Classical, 2004.
Hassler, M., N. Birbaumer, and A. Feil. "Musical talent and visual-spatial abilities: a longitudinal study." *Psychology of Music*, 113 (1985): 99–113.
Mayo Clinic. *Toxic Shock Syndrome*. http://www.mayoclinic.org/diseases-conditions/toxic-shock-syndrome/basics/definition/con-20021326
National Coalition for Music Education, statistics: http://www.oocities.org/wvmea/ncme.htm
National Park Service. *Air Quality at Grand Canyon National Park*. http://nature.nps.gov/air/permits/ARIS/grca/index.cfm
Vasconcelos, Luis A. de P. "Seasonal Transport of Fine Particles to the Grand Canyon." *Journal of the Air and Waste Management Association*, 49 (1999):3, 268–278.
Weinberger, Norman M. "'The Mozart Effect': A Small Part of the Big Picture." *Regents of the University of California*, Vol. VII, Issue 1, Winter 2000.
Williams, Patrick Moody. "Theme for Earth Day." On *The Green Album*, Boston Pops Orchestra, Sony Classical Records, 1992.
Williams, Patrick: his music, http://www.patrickwilliamsmusic.com
Wilson, Frank R., M.D. *Tone Deaf and All Thumbs? An Invitation to Music-Making*. New York: Vintage Books, 1987.
Wong, Lisa M.D. *Scales to Scalpels: Doctors Who Practice the Healing Arts of Music and Medicine: The Story of the Longwood Symphony Orchestra*. New York: Pegasus Books, 2012.

CHAPTER 12

Anderson, Norman D. *The Search for Clean Air, Educational Supplement for the Video*. Chapel Hill, N.C.: North Carolina State University, 1995.
Joseph Kahn and Jim Yardley. "As China Roars, Pollution Reaches Deadly Extremes." *New York Times*, August 26, 2007.
Paul Sisson. "Disease's Spread Traced: Local research shows wind sends Kawasaki from China." *San Diego Union Tribune*, May 20, 2014.
Brimblecombe, Peter (editor), et al. *Acid Rain: Deposition to Recovery*. Springer, 2007.
Burns, J. C., L. Herzog, O. Fabri, A. H. Tremoulet, X. Rodo, R. Uehara, D. Burgner, E. Bainto, D. Pierce, M. Tyree, and D. Cayan. "Seasonality of Kawasaki Disease: A Global Perspective." *Plos One* 8 (2013).

European Environment Agency. "Effects of air pollution on European ecosystems." European Environment Agency Technical Report, 2014.

Hardy, Michael C. *Grandfather Mountain* (Images of America series). Mount Pleasant, S.C.: Arcadia Publishing, 2014.

Morton, Hugh M. *Hugh Morton, North Carolina Photographer.* Chapel Hill, N.C.: University of North Carolina Press, 2006.

Public Broadcasting Service. *The Search for Clean Air.* A Public Broadcasting special film, narrated by Walter Cronkite, 1994.

Rodo, X., J. Ballester, D. Cayan, M. E. Melish, Y. Nakamura, R. Uehara, and J. C. Burns. "Association of Kawasaki disease with tropospheric wind patterns." *Scientific Reports*, 1 (2011).

CHAPTER 13

Aguado, Edward and James E. Burt. *Understanding Weather and Climate.* Upper Saddle River, N.J.: Prentice-Hall, Inc., 1999.

Burns, Robert. *The Complete Poems and Songs of Robert Burns.* Waverly Books Ltd., 2012.

Fujita, T. Theodore. *U.S. Tornadoes Part 1: 70-Year Statistics.* Chicago: University of Chicago Press, 1987.

McDonald, James R. "T. Theodore Fujita: His Contribution to Tornado Knowledge through Damage Documentation and the Fujita Scale." *Bulletin of the American Meteorological Society*, 82(1), January 2001.

Rinehart, Ronald E. *Radar For Meteorologists*, 4th Edition. Rinehart Publications, 2004.

Stocker, Thomas F. "The Silent Services of the World Ocean." *Science*, 350, 6262, November 13, 2015, 764–765.

U.S. Department of Commerce. *Thunderstorms, Tornadoes, Lightning . . . Nature's Most Violent Storms. A Preparedness Guide: Including Safety Information for Schools.* U.S. Department of Commerce, NOAA/PA 201051. www.weather.gov/safety.php

CHAPTER 14

Alexander, M. J., M. Geller, C. McLandress, S. Polavarapu, P. Preusse, F. Sassi, K. Sato, S. Eckermann, M. Ern, A. Hertzog, Y. Kawatani, M. Pulido, T. Shaw, M. Sigmond, R. Vincent, and S. Watanabe. "Recent Developments on Gravity Wave Effects in Climate Models, and the Global Distribution of Gravity Wave Momentum Flux from Observations and Models." *Quarterly Journal of the Royal Meteorological Society*, 136 (2010): 1103–1124.

Allison, Edward H. and Hannah R. Bassett. "Climate Change and the Oceans: Human Impacts and Responses." *Science*, 350, 6262, November 13, 2015, 778–782.

Agence France-Presse in Beijing. "Top Meteorologist Zheng Guoguang Warns of Climate Change Risk to China." *South China Morning Post*, March 22, 2015.

Chen, B., C. Hong, and H. Kan. "Exposures and health outcomes from outdoor air pollutants in China." *Toxicology* (2004) 198(1–3): 291–300.

Climate Change 2001, Intergovernmental Panel on Climate Change (IPCC) Third Assessment Report, World Meteorological Organization and the United Nations Environmental Program.

Climate Change 2004, Intergovernmental Panel on Climate Change (IPCC) Fifth Assessment Report, World Meteorological Organization and the United Nations Environmental Program.

Environmental Protection Agency Household Carbon Footprint Calculator: http://www3.epa.gov/carbon-footprint-calculator

Geller, M. A., M. J. Alexander, P. T. Love, J. Bacmeister, M. Ern, A. Hertzog, E. Manzini, P. Preusse, K. Sato, A. A. Scaife, and T. Zhou. "A Comparison Between Gravity Wave Momentum Fluxes in Observations and Climate Models." *Journal of Climate*, 26 (2013) 17: 6383–6405.

Henson, Robert. *The Thinking Person's Guide to Climate Change*. American Meteorological Society, 2014.

Lane, L. J., M. H. Nichols, and H. B. Osborn. "Time series analyses of global change data." *Environmental Pollution*, 83(1994): 63–68.

NASA Earth Observatory. http://earthobservatory.nasa.gov/

Neuman, Scott. "Top Beijing Scientist: China Faces 'Huge Impact' from Climate Change." National Public Radio, March 22, 2015.

Nevin, Lisa A. and Nadine Le Bris. "The Deep Ocean Under Climate Change." *Science*, 350, 6262, November 13, 2015, 766–768.

Schneider, Stephen H. "On the Carbon Dioxide–Climate Confusion." *Journal of Atmospheric Science*, 32(1975): 2060–2066.

Stern, Arthur et al. *Fundamentals of Air Pollution*, 2nd Edition. Cambridge: Academic Press, Inc., 1984.

Technische Universität Darmstadt, ScienceDaily. "Climate Change: Scientist Investigates Changing Sea Levels." *ScienceDaily*, March 17, 2015.

Wong, C. M., N. Vichit-Vadakan, H. Kan, and Z. Qian. "Public Health and Air Pollution in Asia (PAPA): a multicity study of short-term effects of air pollution on mortality." *Environmental Health Perspectives* 116 (2008): 1195–1202.

Wong, Edward. "On a Scale of 0 to 500, Beijing's Air Quality Tops 'Crazy Bad' at 755." *The New York Times*, January 13, 2013, p. 16.

CHAPTER 15

Airline Ratings: http://www.airlineratings.com/index.php

Federal Aviation Administration. *NASDAC Review of NTSB Weather-Related Accidents, 1993–2003*. Federal Aviation Administration Aviation Safety Information Analysis and Sharing (ASIAS), February 2, 2010. http://www.asias.faa.gov/

Federal Aviation Administration. *NTSB Weather Accidents 1994–2003*. Federal Aviation Administration Aviation Safety Information Analysis and Sharing (ASIAS). http://www.asias.faa.gov/

Federal Aviation Administration. *NTSB Weather Accidents*. Federal Aviation Administration Aviation Safety Information Analysis and Sharing (ASIAS). http://www.asias.faa.gov/

Federal Aviation Administration. *Review of Aviation Accidents Involving Turbulence in the United States 1992–2001*. Federal Aviation Administration, National Aviation Safety Data Analysis Center, FAA Office of System Safety, reference number: 04-551, August 2004.

Lester, Peter F. *Aviation Weather*. Jeppesen Sanderson, Inc., 1997.

CHAPTER 16

Barnes, R. D. *Invertebrate Zoology; Fifth Edition*. Fort Worth, TX: Harcourt Brace Jovanovich College Publishers, 1987. pp. 92–96, 127–134, 149–162.
Buchheim, Jason. *Coral Reef Bleaching*. Odyssey Expeditions—Marine Biology Learning Center Publications, 2013.
Graham, N.A.J., et al. "Climate warming, marine protected areas and the ocean-scale integrity of coral reef ecosystems." *PLOS One* 3 (8), e3039.
Graham, N.A.J., et al. "Dynamic fragility of oceanic coral reef ecosystems." *Proceedings of the National Academy of Sciences* 103 (22), 8425–8429.
Holmes, Sr., Oliver Wendell. *The Complete Poetical Works of Oliver Wendell Holmes*. Cambridge: Houghton Mifflin Company, The Riverside Press, 1910.
McCauley, Douglas J. et al. "Marine defaunation: Animal loss in the global ocean." *Science* 347 (6219), January 16, 2015.
NOAA. *How Do Coral Reefs Form?* http://oceanservice.noaa.gov/education/kits/corals/coral04_reefs.html
Spalding, Mark D. and Barbara E. Brown. "Warm-Water Coral Reefs and Climate Change." *Science* 350 (6262), November 13, 2015, 769–771.

CHAPTER 17

Climate Change Impacts in the United States. U.S. National Climate Assessment, U.S. Global Change Research Program, U.S. Government Printing Office, 2015.
Durran, Dale R. and Dargon M. W. Frierson. "Condensation, Atmospheric Motion, and Cold Beer." *Physics Today*, April 2013, 74–75.
Goklany, Indur M. *Death and Death Rates Due to Extreme Weather Events; Global and U.S. Trends, 1900–2006*. International Policy Press, 2007.
Gramling, Carolyn. "How Warming Oceans Unleashed an Ice Stream." *Science*, 350 (6262), November 13, 2015, 728.
Halverson, Jeffrey B. and Thomas Rabenhorst. "Hurricane Sandy: The Science and Impacts of a Superstorm." *Weatherwise*, March/April 2013, 14–23.
Journal of the Royal Geographic Society, Vol. 50. John Murray, William Clowes & Sons Limited, London, June 1881.
Kay, Jennifer. "Hurricane Center Learns Lessons from Sandy; Changes Warnings." *Insurance Journal*, April 8, 2013.
Landsea, Christopher W. "Hurricanes and Global Warming. Opinion Piece." November 2011. http://www.aoml.noaa.gov/hrd/Landsea/gw_hurricanes/
National Weather Service. "Summary of Natural Hazard Statistics for 2013 in the United States." *National Weather Service Report*, July 22, 2014.
National Weather Service. *74-Year List of Severe Weather Fatalities*. http://www.nws.noaa.gov/om/hazstats/resources/weather_fatalities.pdf
Smith, Adam B. and Richard W. Katz. "U.S. Billion-Dollar Weather and Climate Disasters: Data Sources, Trends, Accuracy and Biases." *Natural Hazards*, 67(2), June 2013, 387–410.
Tallal, Jimy. "Broad Beach in National Spotlight." *Malibu Times*, December 5, 2012.

CHAPTER 18

Corton, Christine L. *London Fog: The Biography*. Belknap Press, 2015.
Cox, William A. M.D. *Early Postmortem Changes and Time of Death*. William A. Cox,

M.D., Forensic Pathologist/Neuropathologist, Dec. 22, 2009. forensicmd.files.wordpress.com.

Cruz, Albert. M. "Crime Scene Intelligence: An Experiment in Forensic Entomology." Occasional Paper Number Twelve, National Defense Intelligence College, Washington, D.C., November 2006.

Gennard. Dorothy E. *Forensic Entomology: An Introduction.* University of Lincoln, United Kingdom, John Wiley & Sons, Ltd., 2007.

Lord, W. D., R. W. Johnson, and F. Johnston. "The Blue Bottle Fly, Calliphora Vicina (*Erythrocephala*) as an Indicator of Human Post-Mortem Interval: A Case of Homicide from Suburban Washington, D.C." *Bull. Soc. Vector Ecol.*, 11(2): 276–280, December 1986.

MacMaster, Greg. *Environmental Forensics: How the Atmosphere Affects Criminal Investigations & Other Professional Research.* Greg MacMaster, 2009.

Marsa, Linda. "Breathless." *Discover*, July 8, 2013, 79–85.

Nash, Linda. *Inescapable Ecologies: A History of Environment, Disease, and Knowledge.* University of California Press, 2006.

National Transportation Safety Board. *Parasailing Safety.* Marine Special Investigative Report. National Transportation Safety Board, NTSB/SIR-14/02 PB2014-106341, 2014.

Parasail Safety Council: http://www.parasail.org/

Pillon, Dennis. "Florida Parasail Regulations Go into Effect, Bill Sponsor Celebrates by Going Parasailing." AL.com, October 3, 2014.

Report on the Accident to Boeing MD11 B-150 at Hong Kong International Airport on 22 August 1999, Aircraft Accident Report 1/2004, Accident Investigation Division, Civil Aviation Department, Hong Kong, December 2004.

Sandburg, Carl. *Carl Sandburg Reads: Grammy Nominee, Fog, the People Yes and Other of His Poems.* Caedmon Audio Cassette, February 1992.

StÆrkeby, Morten. *What Happens After Death?*, University of Oslo, Department of Entomology, 2007.

White House. *The Health Impacts of Climate Change on Americans.* The White House, Washington, D.C., www.whitehouse.gov

Whitman, Edmund. *The Essence of Astronomy: Things Every One Should Know About the Sun, Moon, and Stars.* Scholar's Choice Edition. Scholar's Choice, 2015.

CHAPTER 19

Brown, Norman O., Translator, Hesiod, *Hesiod: Theogony.* Pearson, 1953.

Chambers, R., A. Pacey, and L. A. Thrupp, eds. "Farmer innovation and agricultural research." *Intermediate Technology Publications*, 218, 1989.

Gelbart, Larry, "Gulliver." Lyrics published by Hopeful Music.

Hesiod. "Theogony." A poem. 700 B.C.

Klenk, Nicole L. et al. "Stakeholders in Climate Science: Beyond Lip Service?" *Science* 350 6262), November 13, 2015, 743–744.

Lovelock, James. *Gaia: A New Look at Life on Earth.* Oxford Paperbacks, subsequent edition, 2000.

McIntyre, Beverly D. et al. *Agriculture at a Crossroads—Global Report.* International Assessment of Agricultural Knowledge, Science and Technology for Development, 2009.

Miller, Arthur I. *Jung, Pauli, and the Pursuit of a Scientific Obsession.* W.W. Norton & Company, 2009.

Ooi, Peter A. C. "Bringing Science to Farmers: Experiences in Integrated Diamondback Moth Management." Lead Paper 4, *Proceedings: The Management of Diamondback Moth and Other Crucifer Pests*, 1996, 25–33.

Plastic Paradise: The Great Pacific Garbage Patch. Directed by Angela Sun. An independent documentary film.

Todd, Kim. *Tinkering with Eden.* W.W. Norton & Company, reprinted 2012.

Von Zastrow, Claus. *"If You Protect the Ocean, You Protect Yourself." A Conversation with Jean-Michael Cousteau. LFA: Join the Conversation.* Public School Insights, 2009.

Williams, Patrick Moody. "A Dream That Only I Can Know." Music and lyrics by Patrick Moody Williams. From *Yesterday's Children*, CBS Television Movie, 2000 (Emmy Award winner).

Williams, Patrick Moody. *Gulliver*, featuring the London Royal Philharmonic Orchestra. Composed and conducted by Patrick Moody Williams, narration by Larry Gelbart and read by Sir John Gielgud. Soundwings distributed by The Welk Record Group, 1986 (Grammy Award winner).

Glossary

ALGOR MORTIS: A termed used to describe postmortem cooling of the body.

ATMOSPHERIC AEROSOLS: Aerosols are minute particles suspended in the atmosphere. They interact with the Earth's radiation and climate. They can modify cloud particles, change the clouds' reflective characteristics, including how they absorb sunlight, and have a direct effect on the Earth's energy dispersal.

ATTENUATION: The reduction or weakening of signal strength, whether digital or analog. In meteorology, attenuation of weather radar is caused by the energy of the beam being lost to scattering or absorption; the longer the distance the beam travels, the more hydrometeors (i.e., rain, snow, hail, and such particles) the beam must pass through, increasing the likelihood of attenuation.

BERGERON PROCESS: Also called the Wegener-Bergeron-Findeisen process. A process that explains how precipitation particles form within a cloud composed of both ice crystals and liquid water drops. Since the equilibrium vapor pressure of water vapor over ice is less than over water, once ice crystals begin to form in a cloud, they grow at the expense of the liquid water drops in the cloud.

COFFIN CORNER: The coffin corner (or "Q corner") is the altitude at or near which a fast fixed-wing aircraft's **STALL SPEED** is equal to the critical Mach number, at a given gross weight and G-force loading. At this altitude it is very difficult to keep the airplane in stable flight. Because the stall speed is the minimum speed required to maintain level flight, any reduction in speed will cause the airplane to stall and lose altitude. Because the critical Mach number is the maximum speed at which air can travel over the wings without losing lift due to flow separation and shock waves, any increase in speed will cause the airplane to lose lift, or to pitch heavily nose-down, and lose altitude. The "corner" refers to the triangular shape at the top right of a flight envelope chart where the stall speed and critical Mach number lines come together.

CORIOLIS FORCE: In atmospheric science, an apparent deflection of air from its path as seen by an observer on Earth due to the Earth's rotation.

DEWPOINT TEMPERATURE: The temperature to which air must be cooled, at a certain water vapor content and atmospheric pressure, to reach saturation.

DOBSON UNIT: A measure of **OZONE** concentration. One Dobson Unit is the number of molecules of ozone that would be required to create a layer of pure ozone 0.01 millimeters thick at a temperature of 0 degrees Celsius and a pressure of 1 atmosphere (the air pressure at the surface of the Earth).

EF SCALE: Enhanced Fujita Scale. For tornados, the EF Scale is used to assign a rating based on wind speed and related damage from a tornado.

FUJITA SCALE			DERIVED EF SCALE		OPERATIONAL EF SCALE	
F Number	Fastest 1/4-Mile (mph)	3 Second Gust (mph)	EF Number	3 Second Gust (mph)	EF Number	3 Second Gust (mph)
0	40-72	45-78	0	65-85	0	65-85
1	73-112	79-117	1	86-109	1	86-110
2	113-157	118-161	2	110-137	2	111-135
3	158-207	162-209	3	138-167	3	136-165
4	208-260	210-261	4	168-199	4	166-200
5	261-318	262-317	5	200-234	5	Over 200

*****IMPORTANT NOTE ABOUT ENHANCED F-SCALE WINDS:** *The enhanced F-scale still is a set of wind estimates (not measurements) based on damage.* It uses three-second gusts estimated at the point of damage based on a judgment of 8 levels of damage to the 28 indicators listed below. These estimates vary with height and exposure. **Important:** The 3 second gust is not the same wind as in standard surface observations. Standard measurements are taken by weather stations in open exposures, using a directly measured, "one minute mile" speed.

Source: Storm Prediction Center, NOAA/National Weather Service
Table created by WeatherExtreme Ltd.

FIRNSPIEGEL: *Firn* means "old snow" and *spiegel* means "mirror" in German. Thus the term firnspiegel is often referred to as "mirror snow." It is a common type of snow surface that forms in the springtime; it is a highly reflective snow surface that shines like a mirror when the sun hits it. It forms when the subsurface of the snow melts as the solar energy in the spring penetrates into the snowpack and then the surface of the snow refreezes overnight, forming a thin surface layer on the snowpack.

FRACTAL: From the Latin *fractus*, meaning "irregular." A geometric figure or shape, each part of which has the same statistical character as the whole. Used in modeling structures such as snowflakes, crystal growth, and coastlines, where similar patterns recur at progressively smaller scales. The term was coined by French mathematician Benoit Mandelbrot.

FUJITA SCALE: Scale of tornadic intensity, based on wind speed and damage. (see *EF scale* above)

SCALE	WIND ESTIMATE*** (mph)	TYPICAL DAMAGE
F0	<70	Light Damage. Some damage to chimneys; branches broken off of trees; shallow-rooted trees pushed over; sign boards damaged.
F1	73-112	Moderate Damage. Peels surface off roofs; mobile homes pushed off of foundations or overturned; moving autos blown off roads.
F2	113-157	Considerable Damage. Roofs torn off frame houses; mobile homes demolished; boxcars overturned; large trees snapped or uprooted; light-object missiles generated; cars lifted off ground.
F3	158-206	Severe Damage. Roofs and some walls torn off well-constructed houses; trains overturned; most trees in forest uprooted; heavy cars lifted off the ground and thrown.
F4	207-260	Devastating Damage. Well constructed houses leveled; structures with weak foundations blown away some distance; cars thrown and large missiles generated.
F5	261-318	Incredible Damage. Strong frame houses leveled off foundations and swept away; automobile-sized missiles fly through the air in excess of 100 meters (109 yards); trees debarked; incredible phenomena will occur.

*****IMPORTANT NOTE ABOUT F-SCALE WINDS: Do not use F-scale winds literally. These precise wind speed numbers are actually guesses and have never been scientifically verified. Different wind speeds may cause similar-looking damage from place to place – even from building to building. *Without a thorough engineering analysis of tornado damage in any event, the actual wind speeds needed to cause that damage are unknown.* The Enhanced F-scale was implemented in February 2007.**

Source: Storm Prediction Center, NOAA/National Weather Service

Table created by WeatherExtreme Ltd.

G-FORCES: Gravitational force—a measurement of the force of gravitational acceleration on a particular celestial body. It is measured in g's, where 1 g equals the force of gravity on the Earth's surface. Astronauts in space experience 0 g's while a roller coaster rider might experience 3 g's during the ride.

GALE: An area of low pressure with sustained surface winds of 34 knots (39 mph) to 47 knots (54 mph).

GREENHOUSE GASES: Examples of these are: water vapor, carbon dioxide, ozone, and methane. Heat trapping by these gases is instrumental in controlling the earth's surface temperature. Water vapor is the most abundant greenhouse gas and tends to increase in concentration in response to increased concentration of other greenhouse gases.

LATENT HEAT: Energy absorbed or released by a substance during a change in its physical state while its temperature remains constant.

MACH LIMIT: The limiting speed of a particular airfoil (such as an aircraft wing), beyond which it will potentially come apart. The aerodynamic characteristics of an airfoil change radically when the airfoil exceeds the limiting speed for which it was designed.

MAJOR HURRICANE: Any hurricane classified as category 3 or higher.

MESOCYCLONE: A storm region of rotation of approximately 2–5 miles in diameter and often imbedded in the rear of a supercell or in the front of a high-precipitation storm cell. The area of mesocyclones typically covers a much larger area than that of tornadoes that develop within them. *Mesocyclone* is a radar term that is defined by specific criteria such as magnitude, vertical depth, and duration.

MESOSPHERE: The mesosphere lies directly above the **STRATOSPHERE** and extends from about 50 km (31 miles) to 85 km (53 miles) above the Earth's surface. It is where the coldest temperatures in the Earth's atmosphere are found (near the top of the mesosphere).

MICROBURST: A short downburst of sinking air that covers less that 2.5 miles in scale and which has peak winds lasting 2–5 minutes. A severe microburst can produce winds as high as 150 mph.

MID-LATITUDE CYCLONE: A low-pressure system that has counterclockwise (in the northern hemisphere) flow in the middle latitudes (the latitude belt roughly between 35 and 65 degrees north and south).

MILLIBAR (also **HECTOPASCAL**): A unit of atmospheric pressure. The standard atmospheric pressure at sea level is 1,013.2 millibars.

NOR'EASTER: A nor'easter is a storm that moves along the northeast coast of the United States. The winds in these storms generally blow from the northeast—hence the name. These storms are most likely to occur between September and April.

OZONE: Ozone is a gaseous form of oxygen (chemical formula O_3) that is considered a pollutant at lower altitudes but is a very important chemical in the **STRATOSPHERE**, where it offers us protection from dangerous high-energy radiation from the sun.

POLAR NIGHT JET: This is a stream of strong westerly winds that develop, usually during winter, around 50° to 60° latitude and at about 80,000 feet (24 km). Inside the polar night jet is the **POLAR VORTEX**. The edge of the polar vortex is associated with the core of the polar night jet.

POLAR VORTEX: A large-scale area of low pressure and very cold air at both of the Earth's poles that is contained by the **POLAR NIGHT JET**. The polar vortex extends from the upper **TROPOSPHERE** around the **TROPOPAUSE** region through the **STRATOSPHERE** and into the **MESOSPHERE**.

RIME: Rime is a milky, granular, and opaque form of ice that is formed by the freezing of **SUPERCOOLED** water drops as they impact an object like a the wing of an aircraft.

SAFFIR-SIMPSON HURRICANE WIND SCALE: This is a wind rating scale that goes from 1 to 5, based on the sustained winds of a hurricane. It basically goes from "very dangerous" to "catastrophic." Category 1 wind ranges are 74–95 mph; category 2, 90–110 mph; category 3, 111–129 mph; category 4, 130–156 mph; and category 5, 157 mph or higher.

SAINT FRANCIS OF ASSISI: A saint who was, among other things, famous for his love and caring for animals. He was canonized in 1228 and is the patron saint of animals and the environment.

SOLAR WIND: This is a flow of ionized gas, mainly composed of hydrogen, that flows continuously outward from the sun. It flows at a very high velocity and can vary in intensity. The solar wind is not uniform and is always directed away from the sun. If it passes near the Earth's magnetic field, it can produce some atmospheric effects, such as auroras.

STALL SPEED: A condition in aerodynamics in which the angle of attack of an airfoil increases beyond a certain point, at which the lift begins to decrease and the airfoil is no longer "flying"—it stalls.

STRATOSPHERE: The second layer in the Earth's atmosphere as you go upward. It extends from the top of the **TROPOSPHERE** to the bottom of the **MESOSPHERE**. The height of the bottom of the stratosphere varies with latitude and seasons. Unlike in the troposphere, air temperature increases as you go upward into the stratosphere. It is an important area in the study and measurement of **OZONE**.

SUNSPOTS: These are dark regions on the sun that are the result of intense magnetic activity, solar flares, and coronal mass ejections. They appear dark because they are considerably cooler than the rest of the surface of the sun.

SUPERCELL: These are severe single-cell thunderstorms. They are the least common, but can be extremely destructive when they occur. They produce deep rotating updrafts called **MESOCYCLONES** and can last several hours.

SUPERCOOL: When a liquid gets below its normal freezing point without being solid or crystalized.

THUNDERSTORM: A storm made up of cumulonimbus clouds and producing lightning and thunder.

TROPOPAUSE: This is the upper boundary of the troposphere. At this level the decreasing temperature with height usually neutralizes and is more constant with altitude.

TROPOSPHERE: The layer of the atmosphere that starts at the Earth's surface and extends up to the **TROPOPAUSE**. The temperature of the air in the troposphere decreases with altitude.

WET BULB TEMPERATURE: This is the lowest temperature of an air parcel if it were cooled adiabatically (i.e., without gain or loss of heat) at constant pressure to saturation by the evaporation of water into it.

WIND SHEAR: The rate at which the velocity of wind changes from point A to point B in a given direction (horizontally or vertically).

WORLD WAR II ALLIED POWERS: United States, Britain, France, USSR (Union of Soviet Socialist Republics), Australia, Belgium, Brazil, Canada, China, Denmark, Greece, the Netherlands, New Zealand, Norway, Poland, South Africa, Yugoslavia.

WORLD WAR II AXIS POWERS: Germany, Italy, Japan, Hungary, Romania, Bulgaria.

Index

CPSIA information can be obtained
at www.ICGtesting.com
Printed in the USA
LVHW041754130521
687356LV00011B/1243

9 781681 774039